51CTO学院丛书

Flink
入门与实战

徐葳 著

人民邮电出版社
北京

图书在版编目（CIP）数据

Flink入门与实战 / 徐葳著. —— 北京：人民邮电出版社，2019.10
（51CTO学院丛书）
ISBN 978-7-115-51678-7

Ⅰ．①F… Ⅱ．①徐… Ⅲ．①数据处理软件 Ⅳ．①TP274

中国版本图书馆CIP数据核字(2019)第192805号

内 容 提 要

　　本书旨在帮助读者从零开始快速掌握Flink的基本原理与核心功能。本书首先介绍了Flink的基本原理和安装部署，并对Flink中的一些核心API进行了详细分析。然后配套对应的案例分析，分别使用Java代码和Scala代码实现案例。最后通过两个项目演示了Flink在实际工作中的一些应用场景，帮助读者快速掌握Flink开发。

　　学习本书需要大家具备一些大数据的基础知识，比如Hadoop、Kafka、Redis、Elasticsearch等框架的基本安装和使用。本书也适合对大数据实时计算感兴趣的读者阅读。

◆ 著　　　　徐 葳
　　责任编辑　陈聪聪
　　责任印制　焦志炜

◆ 人民邮电出版社出版发行　北京市丰台区成寿寺路11号
　　邮编　100164　电子邮件　315@ptpress.com.cn
　　网址　https://www.ptpress.com.cn
　　北京七彩京通数码快印有限公司印刷

◆ 开本：800×1000　1/16
　　印张：15　　　　　　　　　2019年10月第1版
　　字数：264千字　　　　　　2025年1月北京第14次印刷

定价：59.00 元

读者服务热线：(010)81055410　印装质量热线：(010)81055316
反盗版热线：(010)81055315
广告经营许可证：京东市监广登字20170147号

前言

Flink项目是大数据计算领域冉冉升起的一颗新星。大数据计算引擎的发展经历了几个过程，从第1代的MapReduce，到第2代基于有向无环图的Tez，第3代基于内存计算的Spark，再到第4代的Flink。因为Flink可以基于Hadoop进行开发和使用，所以Flink并不会取代Hadoop，而是和Hadoop紧密结合。

Flink主要包括DataStream API、DataSet API、Table API、SQL、Graph API和FlinkML等。现在Flink也有自己的生态圈，涉及离线数据处理、实时数据处理、SQL操作、图计算和机器学习库等。

本书共分11章，每章的主要内容如下。

第1~2章主要针对Flink的原理组件进行分析，其中包括针对Storm、Spark Streaming和Flink这3个实时计算框架进行对比和分析，以及快速实现Flink的入门案例开发。

第3章主要介绍Flink的安装部署，包含Flink的几种部署模式：本地模式、Standalone模式和YARN模式。本章主要针对YARN模式进行了详细分析，因为在实际工作中以YARN模式为主，这样可以充分利用现有集群资源。

第4章主要针对DataStream和DataSet中的相关组件及API进行详细分析，并对Table API和SQL操作进行了基本的分析。

第5~9章主要针对Flink的一些高级特性进行了详细的分析，包含Broadcast、Accumulator、Distributed Cache、State、CheckPoint、StateBackend、SavePoint、Window、Time、Watermark以及Flink中的并行度。

第10章主要介绍常用组件Kafka-Connector，针对Kafka Consumer和Kafka Producer的使用结合具体案例进行分析，并描述了Kafka的容错机制的应用。

第11章介绍Flink在实际工作中的两个应用场景：一个是实时数据清洗(实时ETL)，另一个是实时数据报表，通过这两个项目实战可以加深对Flink的理解。

感谢所有在本书编写过程中提出宝贵意见的朋友。作者水平有限，书中如有不足之处还望指出并反馈至邮箱xuwei@xuwei.tech，作者将不胜感激。

资源与支持

本书由异步社区出品，社区（https://www.epubit.com/）为您提供相关资源和后续服务。

配套资源

本书提供如下资源：

本书配套资源请到异步社区本书购买页处下载。

要获得以上配套资源，请在异步社区本书页面中单击 配套资源 ，跳转到下载界面，按提示进行操作即可。注意：为保证购书读者的权益，该操作会给出相关提示，要求输入提取码进行验证。

提交勘误

作者和编辑尽最大努力来确保书中内容的准确性，但难免会存在疏漏。欢迎您将发现的问题反馈给我们，帮助我们提升图书的质量。

当您发现错误时，请登录异步社区，按书名搜索，进入本书页面，单击"提交勘误"，输入勘误信息，单击"提交"按钮即可（见下图）。本书的作者和编辑会对您提交的勘误进行审核，确认并接受后，您将获赠异步社区的100积分。积分可用于在异步社区兑换优惠券、样书或奖品。

与我们联系

我们的联系邮箱是 contact@epubit.com.cn。

如果您对本书有任何疑问或建议,请您发邮件给我们,并请在邮件标题中注明本书书名,以便我们更高效地做出反馈。

如果您有兴趣出版图书、录制教学视频,或者参与图书翻译、技术审校等工作,可以发邮件给我们;有意出版图书的作者也可以到异步社区在线提交投稿(直接访问www.epubit.com/selfpublish/submission即可)。

如果所在的学校、培训机构或企业想批量购买本书,或异步社区出版的其他图书,也可以发邮件给我们。

如果您在网上发现有针对异步社区出品图书的各种形式的盗版行为,包括对图书全部或部分内容的非授权传播,请您将怀疑有侵权行为的链接发邮件给我们。您的这一举动是对作者权益的保护,也是我们持续为您提供有价值的内容的动力之源。

关于异步社区和异步图书

"异步社区"是人民邮电出版社旗下IT专业图书社区,致力于出版精品IT技术图书和相关学习产品,为作译者提供优质出版服务。异步社区创办于2015年8月,提供大量精品IT技术图书和电子书,以及高品质技术文章和视频课程。更多详情请访问异步社区官网 https://www.epubit.com。

"异步图书"是由异步社区编辑团队策划出版的精品IT专业图书的品牌,依托于人民邮电出版社近30年的计算机图书出版积累和专业编辑团队,相关图书在封面上印有异步图书的LOGO。异步图书的出版领域包括软件开发、大数据、AI、测试、前端、网络技术等。

异步社区

微信服务号

目录

第1章 Flink概述 ... 1
1.1 Flink原理分析 ... 1
1.2 Flink架构分析 ... 2
1.3 Flink基本组件 ... 3
1.4 Flink流处理（Streaming）与批处理（Batch） ... 4
1.5 Flink典型应用场景分析 ... 5
1.6 流式计算框架对比 ... 6
1.7 工作中如何选择实时计算框架 ... 8

第2章 Flink快速入门 ... 9
2.1 Flink开发环境分析 ... 9
    2.1.1 开发工具推荐 ... 9
    2.1.2 Flink程序依赖配置 ... 10
2.2 Flink程序开发步骤 ... 11
2.3 Flink流处理（Streaming）案例开发 ... 11
    2.3.1 Java代码开发 ... 12
    2.3.2 Scala代码开发 ... 14
    2.3.3 执行程序 ... 16
2.4 Flink批处理（Batch）案例开发 ... 16
    2.4.1 Java代码开发 ... 16
    2.4.2 Scala代码开发 ... 18

2.4.3 执行程序 ······ 19

第3章 Flink的安装和部署 ······ 20

3.1 Flink本地模式 ······ 20
3.2 Flink集群模式 ······ 22
    3.2.1 Standalone模式 ······ 23
    3.2.2 Flink on Yarn模式 ······ 26
    3.2.3 yarn-session.sh命令分析 ······ 30
    3.2.4 Flink run命令分析 ······ 30
3.3 Flink代码生成JAR包 ······ 31
3.4 Flink HA的介绍和使用 ······ 35
    3.4.1 Flink HA ······ 35
    3.4.2 Flink Standalone集群的HA安装和配置 ······ 35
    3.4.3 Flink on Yarn集群HA的安装和配置 ······ 50
3.5 Flink Scala Shell ······ 53

第4章 Flink常用API详解 ······ 56

4.1 Flink API的抽象级别分析 ······ 56
4.2 Flink DataStream的常用API ······ 57
    4.2.1 DataSource ······ 57
    4.2.2 Transformation ······ 66
    4.2.3 Sink ······ 70
4.3 Flink DataSet的常用API分析 ······ 80
    4.3.1 DataSource ······ 80
    4.3.2 Transformation ······ 81
    4.3.3 Sink ······ 82
4.4 Flink Table API和SQL的分析及使用 ······ 82
    4.4.1 Table API和SQL的基本使用 ······ 83
    4.4.2 DataStream、DataSet和Table之间的转换 ······ 87
    4.4.3 Table API和SQL的案例 ······ 91
4.5 Flink支持的DataType分析 ······ 97
4.6 Flink序列化分析 ······ 97

第 5 章 Flink 高级功能的使用 ·········· 99

- 5.1 Flink Broadcast ·········· 99
- 5.2 Flink Accumulator ·········· 104
- 5.3 Flink Broadcast 和 Accumulator 的区别 ·········· 108
- 5.4 Flink Distributed Cache ·········· 108

第 6 章 Flink State 管理与恢复 ·········· 112

- 6.1 State ·········· 112
  - 6.1.1 Keyed State ·········· 113
  - 6.1.2 Operator State ·········· 115
- 6.2 State 的容错 ·········· 116
- 6.3 CheckPoint ·········· 118
- 6.4 StateBackend ·········· 119
- 6.5 Restart Strategy ·········· 121
- 6.6 SavePoint ·········· 123

第 7 章 Flink 窗口详解 ·········· 125

- 7.1 Window ·········· 125
- 7.2 Window 的使用 ·········· 126
  - 7.2.1 Time Window ·········· 127
  - 7.2.2 Count Window ·········· 128
  - 7.2.3 自定义 Window ·········· 129
- 7.3 Window 聚合分类 ·········· 130
  - 7.3.1 增量聚合 ·········· 130
  - 7.3.2 全量聚合 ·········· 132

第 8 章 Flink Time 详解 ·········· 134

- 8.1 Time ·········· 134
- 8.2 Flink 如何处理乱序数据 ·········· 135
  - 8.2.1 Watermark ·········· 136

目录

	8.2.2 Watermark 的生成方式	137
8.3	EventTime+Watermark 解决乱序数据的案例详解	138
	8.3.1 实现 Watermark 的相关代码	138
	8.3.2 通过数据跟踪 Watermark 的时间	142
	8.3.3 利用 Watermark+Window 处理乱序数据	149
	8.3.4 Late Element 的处理方式	153
	8.3.5 在多并行度下的 Watermark 应用	163
	8.3.6 With Periodic Watermarks 案例总结	165

第 9 章 Flink 并行度详解 ... 166

9.1	Flink 并行度	166
9.2	TaskManager 和 Slot	166
9.3	并行度的设置	167
	9.3.1 并行度设置之 Operator Level	168
	9.3.2 并行度设置之 Execution Environment Level	168
	9.3.3 并行度设置之 Client Level	169
	9.3.4 并行度设置之 System Level	169
9.4	并行度案例分析	169

第 10 章 Flink Kafka Connector 详解 ... 172

10.1	Kafka Connector	172
10.2	Kafka Consumer	173
	10.2.1 Kafka Consumer 消费策略设置	173
	10.2.2 Kafka Consumer 的容错	175
	10.2.3 动态加载 Topic	176
	10.2.4 Kafka Consumer Offset 自动提交	177
10.3	Kafka Producer	177
	10.3.1 Kafka Producer 的使用	177
	10.3.2 Kafka Producer 的容错	179

第11章 Flink实战项目开发 ……184

11.1 实时数据清洗（实时ETL） …… 184
11.1.1 需求分析 …… 184
11.1.2 项目架构设计 …… 184
11.1.3 项目代码实现 …… 186

11.2 实时数据报表 …… 205
11.2.1 需求分析 …… 205
11.2.2 项目架构设计 …… 206
11.2.3 项目代码实现 …… 207

第1章
Flink 概述

本章讲解 Flink 的基本原理，主要包含 Flink 原理及架构分析、Flink 组件介绍、Flink 中的流处理和批处理的对比、Flink 的一些典型应用场景分析，以及 Flink 和其他流式计算框架的区别等。

1.1 Flink 原理分析

很多人是在 2015 年才听到 Flink 这个词的，其实早在 2008 年，Flink 的前身就已经是柏林理工大学的一个研究性项目，在 2014 年这个项目被 Apache 孵化器所接受后，Flink 迅速成为 ASF（Apache Software Foundation）的顶级项目之一。截至目前，Flink 的版本经过了多次更新，本书基于 1.6 版本写作。

Flink 是一个开源的流处理框架，它具有以下特点。

- 分布式：Flink 程序可以运行在多台机器上。
- 高性能：处理性能比较高。
- 高可用：由于 Flink 程序本身是稳定的，因此它支持高可用性（High Availability, HA）。
- 准确：Flink 可以保证数据处理的准确性。

Flink 主要由 Java 代码实现，它同时支持实时流处理和批处理。对于 Flink 而言，作为一个流处理框架，批数据只是流数据的一个极限特例而已。此外，Flink 还支持迭代计算、内存管理和程序优化,这是它的原生特性。

由图1.1可知,Flink的功能特性如下。

- 流式优先:Flink可以连续处理流式数据。
- 容错:Flink提供有状态的计算,可以记录数据的处理状态,当数据处理失败的时候,能够无缝地从失败中恢复,并保持Exactly-once。
- 可伸缩:Flink中的一个集群支持上千个节点。
- 性能:Flink支持高吞吐、低延迟。

在这里解释一下,高吞吐表示单位时间内可以处理的数据量很大,低延迟表示数据产生以后可以在很短的时间内对其进行处理,也就是Flink可以支持快速地处理海量数据。

图1.1 Flink的功能特性

1.2 Flink架构分析

Flink架构可以分为4层,包括Deploy层、Core层、API层和Library层,如图1.2所示。

- Deploy层:该层主要涉及Flink的部署模式,Flink支持多种部署模式——本地、集群(Standalone/YARN)和云服务器(GCE/EC2)。
- Core层:该层提供了支持Flink计算的全部核心实现,为API层提供基础服务。

- API层：该层主要实现了面向无界Stream的流处理和面向Batch的批处理API，其中流处理对应DataStream API，批处理对应DataSet API。
- Library层：该层也被称为Flink应用框架层，根据API层的划分，在API层之上构建的满足特定应用的实现计算框架，也分别对应于面向流处理和面向批处理两类。面向流处理支持CEP（复杂事件处理）、基于SQL-like的操作（基于Table的关系操作）；面向批处理支持FlinkML（机器学习库）、Gelly（图处理）、Table操作。

从图1.2可知，Flink对底层的一些操作进行了封装，为用户提供了DataStream API和DataSet API。使用这些API可以很方便地完成一些流数据处理任务和批数据处理任务。

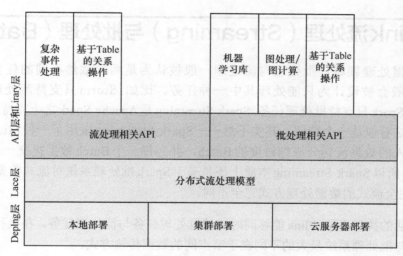

图1.2　Flink架构

1.3　Flink基本组件

读者应该对Hadoop和Storm程序有所了解，在Hadoop中实现一个MapReduce需要两个阶段——Map和Reduce，而在Storm中实现一个Topology则需要Spout和Bolt组件。因此，如果我们想实现一个Flink任务的话，也需要有类似的逻辑。

Flink中提供了3个组件，包括DataSource、Transformation和DataSink。

- DataSource：表示数据源组件，主要用来接收数据，目前官网提供了readTextFile、socketTextStream、fromCollection以及一些第三方的Source。

- Transformation：表示算子，主要用来对数据进行处理，比如Map、FlatMap、Filter、Reduce、Aggregation等。

- DataSink：表示输出组件，主要用来把计算的结果输出到其他存储介质中，比如writeAsText以及Kafka、Redis、Elasticsearch等第三方Sink组件。

因此，想要组装一个Flink Job，至少需要这3个组件。

Flink Job=DataSource+Transformation+DataSink

1.4 Flink流处理（Streaming）与批处理（Batch）

在大数据处理领域，批处理与流处理一般被认为是两种截然不同的任务，一个大数据框架一般会被设计为只能处理其中一种任务。比如，Storm只支持流处理任务，而MapReduce、Spark只支持批处理任务。Spark Streaming是Apache Spark之上支持流处理任务的子系统，这看似是一个特例，其实不然——Spark Streaming采用了一种Micro-Batch架构，即把输入的数据流切分成细粒度的Batch，并为每一个Batch数据提交一个批处理的Spark任务，所以Spark Streaming本质上还是基于Spark批处理系统对流式数据进行处理，和Storm等完全流式的数据处理方式完全不同。

通过灵活的执行引擎，Flink能够同时支持批处理任务与流处理任务。在执行引擎层级，流处理系统与批处理系统最大的不同在于节点间的数据传输方式。

如图1.3所示，对于一个流处理系统，其节点间数据传输的标准模型是，在处理完成一条数据后，将其序列化到缓存中，并立刻通过网络传输到下一个节点，由下一个节点继续处理。而对于一个批处理系统，其节点间数据传输的标准模型是，在处理完成一条数据后，将其序列化到缓存中，当缓存写满时，就持久化到本地硬盘上；在所有数据都被处理完成后，才开始将其通过网络传输到下一个节点。

这两种数据传输模式是两个极端，对应的是流处理系统对低延迟和批处理系统对高吞吐的要求。Flink的执行引擎采用了一种十分灵活的方式，同时支持了这两种数据传输模型。

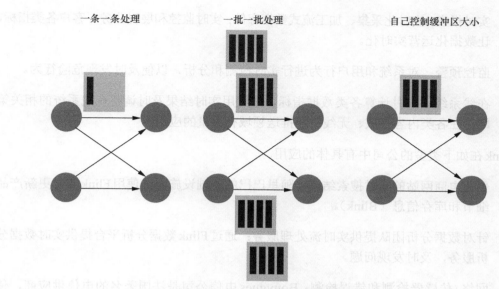

图1.3　Flink的3种数据传输模型

Flink以固定的缓存块为单位进行网络数据传输，用户可以通过设置缓存块超时值指定缓存块的传输时机。如果缓存块的超时值为0，则Flink的数据传输方式类似于前面所提到的流处理系统的标准模型，此时系统可以获得最低的处理延迟；如果缓存块的超时值为无限大，则Flink的数据传输方式类似于前面所提到的批处理系统的标准模型，此时系统可以获得最高的吞吐量。

缓存块的超时值也可以设置为0到无限大之间的任意值，缓存块的超时阈值越小，Flink流处理执行引擎的数据处理延迟就越低，但吞吐量也会降低，反之亦然。通过调整缓存块的超时阈值，用户可根据需求灵活地权衡系统延迟和吞吐量。

1.5　Flink典型应用场景分析

Flink主要应用于流式数据分析场景，目前涉及如下领域。

- 实时ETL：集成流计算现有的诸多数据通道和SQL灵活的加工能力，对流式数据进行实时清洗、归并和结构化处理；同时，对离线数仓进行有效的补充和优化，并为数据实时传输提供可计算通道。

- 实时报表：实时化采集、加工流式数据存储；实时监控和展现业务、客户各类指标，让数据化运营实时化。
- 监控预警：对系统和用户行为进行实时检测和分析，以便及时发现危险行为。
- 在线系统：实时计算各类数据指标，并利用实时结果及时调整在线系统的相关策略，在各类内容投放、无线智能推送领域有大量的应用。

Flink 在如下类型的公司中有具体的应用。

- 优化电商网站的实时搜索结果：阿里巴巴的基础设施团队使用 Flink 实时更新产品细节和库存信息（Blink）。
- 针对数据分析团队提供实时流处理服务：通过 Flink 数据分析平台提供实时数据分析服务，及时发现问题。
- 网络/传感器检测和错误检测：Bouygues 电信公司是法国著名的电信供应商，使用 Flink 监控其有线和无线网络，实现快速故障响应。
- 商业智能分析 ETL：Zalando 使用 Flink 转换数据以便于将其加载到数据仓库，简化复杂的转换操作，并确保分析终端用户可以更快地访问数据（实时 ETL）。

1.6 流式计算框架对比

Storm 是比较早的流式计算框架，后来又出现了 Spark Streaming 和 Trident，现在又出现了 Flink 这种优秀的实时计算框架，那么这几种计算框架到底有什么区别呢？下面我们来详细分析一下，如表 1.1 所示。

表 1.1 流式计算框架对比

产品	模型	API	保证次数	容错机制	状态管理	延时	吞吐量
Storm	Native（数据进入立即处理）	组合式（基础API）	At-least-once（至少一次）	Record ACK（ACK机制）	无	低	低
Trident	Micro-Batching（划分为小批处理）	组合式	Exactly-once（仅一次）	Record ACK	基于操作（每次操作有一个状态）	中等	中等

续表

产品	模型	API	保证次数	容错机制	状态管理	延时	吞吐量
Spark Streaming	Micro-Batching	声明式（提供封装后的高阶函数，如count函数）	Exactly-once	RDD CheckPoint（基于RDD做CheckPoint）	基于DStream	中等	高
Flink	Native	声明式	Exactly-once	CheckPoint（Flink的一种快照）	基于操作	低	高

在这里对这几种框架进行对比。

- 模型：Storm和Flink是真正的一条一条处理数据；而Trident(Storm的封装框架)和Spark Streaming其实都是小批处理，一次处理一批数据（小批量）。

- API：Storm和Trident都使用基础API进行开发，比如实现一个简单的sum求和操作；而Spark Streaming和Flink中都提供封装后的高阶函数，可以直接拿来使用，这样就比较方便了。

- 保证次数：在数据处理方面，Storm可以实现至少处理一次，但不能保证仅处理一次，这样就会导致数据重复处理问题，所以针对计数类的需求，可能会产生一些误差；Trident通过事务可以保证对数据实现仅一次的处理，Spark Streaming和Flink也是如此。

- 容错机制：Storm和Trident可以通过ACK机制实现数据的容错机制，而Spark Streaming和Flink可以通过CheckPoint机制实现容错机制。

- 状态管理：Storm中没有实现状态管理，Spark Streaming实现了基于DStream的状态管理，而Trident和Flink实现了基于操作的状态管理。

- 延时：表示数据处理的延时情况，因此Storm和Flink接收到一条数据就处理一条数据，其数据处理的延时性是很低的；而Trident和Spark Streaming都是小型批处理，它们数据处理的延时性相对会偏高。

- 吞吐量：Storm的吞吐量其实也不低，只是相对于其他几个框架而言较低；Trident属于中等；而Spark Streaming和Flink的吞吐量是比较高的。

官网中Flink和Storm的吞吐量对比如图1.4所示。

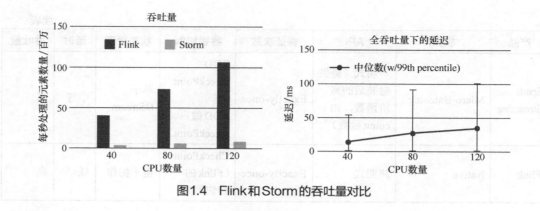

图1.4 Flink和Storm的吞吐量对比

1.7 工作中如何选择实时计算框架

前面我们分析了3种实时计算框架，那么公司在实际操作时到底选择哪种技术框架呢？下面我们来分析一下。

- 需要关注流数据是否需要进行状态管理，如果是，那么只能在Trident、Spark Streaming和Flink中选择一个。
- 需要考虑项目对At-least-once（至少一次）或者Exactly-once（仅一次）消息投递模式是否有特殊要求，如果必须要保证仅一次，也不能选择Storm。
- 对于小型独立的项目，并且需要低延迟的场景，建议使用Storm，这样比较简单。
- 如果你的项目已经使用了Spark，并且秒级别的实时处理可以满足需求的话，建议使用Spark Streaming
- 要求消息投递语义为Exactly-once；数据量较大，要求高吞吐低延迟；需要进行状态管理或窗口统计，这时建议使用Flink。

第2章
Flink快速入门

第1章针对Flink的基本原理、架构和组件进行了分析，本章开始快速实现一个Flink的入门案例，这样可以加深对之前内容的理解。

2.1　Flink开发环境分析

2.1.1　开发工具推荐

在实战之前，需要先说明一下开发工具的问题。官方建议使用IntelliJ IDEA，因为它默认集成了Scala和Maven环境，使用更加方便，当然使用Eclipse也是可以的。

开发Flink程序时，可以使用Java或者Scala语言，个人建议使用Scala，因为使用Scala实现函数式编程会比较简洁。当然使用Java也可以，只不过实现起来代码逻辑比较笨重罢了。

在开发Flink程序的时候，建议使用Maven管理依赖。针对Maven仓库，建议使用国内镜像仓库地址，因为国外仓库下载较慢，可以使用国内阿里云的Maven仓库。

注意：如果发现依赖国内源无法下载的时候，记得切换回国外源。利用国内阿里云Maven仓库镜像进行相关配置时，需要修改 $Maven_HOME/conf/settings.xml 文件。

```
<mirror>
    <id>aliMaven</id>
    <name>aliyun Maven</name>
    <url>http://Maven.aliyun.com/nexus/content/groups/public/</url>
```

```
            <mirrorOf>central</mirrorOf>
    </mirror>
```

2.1.2　Flink程序依赖配置

在使用Maven管理Flink程序相关依赖的时候，需要提前将它们配置好。对应的Maven项目创建完成以后，也需要在这个项目的pom.xml文件中进行相关配置。

使用Java语言开发Flink程序的时候需要添加以下配置。

注意：在这里使用的Flink版本是1.6.1。如果使用的是其他版本，需要到Maven仓库中查找对应版本的Maven配置。

```
<dependency>
    <groupId>org.apache.flink</groupId>
    <artifactId>flink-java</artifactId>
    <version>1.6.1</version>
    <scope>provided</scope>
</dependency>
<dependency>
    <groupId>org.apache.flink</groupId>
    <artifactId>flink-streaming-java_2.11</artifactId>
    <version>1.6.1</version>
    <scope>provided</scope>
</dependency>
```

使用Scala语言开发Flink程序的时候需要添加下面的配置。

```
<dependency>
    <groupId>org.apache.flink</groupId>
    <artifactId>flink-scala_2.11</artifactId>
    <version>1.6.1</version>
    <scope>provided</scope>
</dependency>
<dependency>
    <groupId>org.apache.flink</groupId>
    <artifactId>flink-streaming-scala_2.11</artifactId>
    <version>1.6.1</version>
    <scope>provided</scope>
</dependency>
```

注意：在IDEA等开发工具中运行代码的时候，需要把依赖配置中的scope属性注释掉。在编译打JAR包的时候，需要开启scope属性，这样最终的JAR包就不会把这些依赖包也包含进去，因为集群中本身是有Flink的相关依赖的。

2.2 Flink程序开发步骤

开发Flink程序有固定的流程。

（1）获得一个执行环境。

（2）加载/创建初始化数据。

（3）指定操作数据的Transaction算子。

（4）指定计算好的数据的存放位置。

（5）调用execute()触发执行程序。

注意：Flink程序是延迟计算的，只有最后调用execute()方法的时候才会真正触发执行程序。

延迟计算的好处：你可以开发复杂的程序，Flink会将这个复杂的程序转成一个Plan，并将Plan作为一个整体单元执行！

在这里，提前创建一个Flink的Maven项目，起名为FlinkExample，效果如图2.1所示。

后面的Java代码全部存放在src/main/Java目录下，Scala代码全部存放在src/main/Scala目录下，流计算相关的代码存放在对应的streaming目录下，批处理相关的代码则存放在对应的batch目录下。

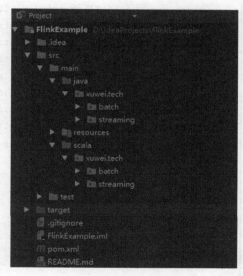

图2.1　项目目录

2.3 Flink流处理（Streaming）案例开发

需求分析：通过Socket手工实时产生一些单词，使用Flink实时接收数据，对指定时间窗口内（如2s）的数据进行聚合统计，并且把时间窗口内计算的结果打印出来。

2.3.1 Java代码开发

首先添加Java代码对应的Maven依赖，参考2.1.2节的内容。注意，在下面的代码中，我们会创建一个WordWithCount类，这个类主要是为了方便统计每个单词出现的总次数。

需求：实现每隔1s对最近2s内的数据进行汇总计算。

分析：通过Socket模拟产生单词，使用Flink程序对数据进行汇总计算。

代码实现如下。

```java
package xuwei.tech.streaming;

import org.apache.Flink.api.common.functions.FlatMapFunction;
import org.apache.Flink.api.Java.utils.ParameterTool;
import org.apache.Flink.contrib.streaming.state.RocksDBStateBackend;
import org.apache.Flink.runtime.state.filesystem.FsStateBackend;
import org.apache.Flink.runtime.state.memory.MemoryStateBackend;
import org.apache.Flink.streaming.api.DataStream.DataStream;
import org.apache.Flink.streaming.api.DataStream.DataStreamSource;
import org.apache.Flink.streaming.api.environment.StreamExecutionEnvironment;
import org.apache.Flink.streaming.api.windowing.time.Time;
import org.apache.Flink.util.Collector;

/**
 * 单词计数之滑动窗口计算
 *
 * Created by xuwei.tech
 */
public class SocketWindowWordCountJava {

    public static void main(String[] args) throws Exception{
        //获取需要的端口号
        int port;
        try {
            ParameterTool parameterTool = ParameterTool.fromArgs(args);
            port = parameterTool.getInt("port");
        }catch (Exception e){
            System.err.println("No port set. use default port 9000--Java");
```

```java
            port = 9000;
        }

        //获取Flink的运行环境
        StreamExecutionEnvironment env = StreamExecutionEnvironment.getExecutionEnvironment();

        String hostname = "hadoop100";
        String delimiter = "\n";
        //连接Socket获取输入的数据
        DataStreamSource<String> text = env.socketTextStream(hostname, port, delimiter);

        // a a c

        // a 1
        // a 1
        // c 1
        DataStream<WordWithCount> windowCounts = text.flatMap(new FlatMapFunction<String, WordWithCount>() {
            public void flatMap(String value, Collector<WordWithCount> out) throws Exception {
                String[] splits = value.split("\\s");
                for (String word : splits) {
                    out.collect(new WordWithCount(word, 1L));
                }
            }
        }).keyBy("word")
                .timeWindow(Time.seconds(2), Time.seconds(1))//指定时间窗口大小为2s,指定时间间隔为1s
                .sum("count");//在这里使用sum或者reduce都可以
                /*.reduce(new ReduceFunction<WordWithCount>() {
                                        public WordWithCount reduce(WordWithCount a, WordWithCount b) throws Exception {

                                            return new WordWithCount(a.word,a.count+b.count);
                                        }
                                    })*/
        //把数据打印到控制台并且设置并行度
        windowCounts.print().setParallelism(1);
        //这一行代码一定要实现,否则程序不执行
```

```
            env.execute("Socket window count");
        }
    }
    public static class WordWithCount{
        public String word;
        public long count;
        public  WordWithCount(){}
        public WordWithCount(String word,long count){
            this.word = word;
            this.count = count;
        }
        @Override
        public String toString() {
            return "WordWithCount{" +
                    "word='" + word + '\'' +
                    ", count=" + count +
                    '}';
        }
    }
}
```

2.3.2 Scala代码开发

首先添加Scala代码对应的Maven依赖，参考2.1.2节的内容。在这里通过case class的方式在Scala中创建一个类。

需求：实现每隔1s对最近2s内的数据进行汇总计算。

分析：通过Socket模拟产生单词，使用Flink程序对数据进行汇总计算。

代码实现如下。

```
package xuwei.tech.streaming

import org.apache.Flink.api.Java.utils.ParameterTool
import org.apache.Flink.streaming.api.Scala.StreamExecutionEnvironment
import org.apache.Flink.streaming.api.windowing.time.Time

/**
 * 单词计数之滑动窗口计算
```

```scala
 *
 * Created by xuwei.tech
 */
object SocketWindowWordCountScala {

  def main(args: Array[String]): Unit = {

    //获取Socket端口号
    val port: Int = try {
      ParameterTool.fromArgs(args).getInt("port")
    }catch {
      case e: Exception => {
        System.err.println("No port set. use default port 9000--Scala")
      }
        9000
    }


    //获取运行环境
    val env: StreamExecutionEnvironment = StreamExecutionEnvironment.getExecutionEnvironment

    //连接Socket获取输入数据
    val text = env.socketTextStream("hadoop100",port,'\n')


    //解析数据（把数据打平），分组，窗口计算，并且聚合求sum

    //注意：必须要添加这一行隐式转行，否则下面的FlatMap方法执行会报错
    import org.apache.Flink.api.Scala._

    val windowCounts = text.flatMap(line => line.split("\\s"))//打平，把每一行单词都切开
      .map(w => WordWithCount(w,1))//把单词转成word , 1这种形式
      .keyBy("word")//分组
      .timeWindow(Time.seconds(2),Time.seconds(1))//指定窗口大小,指定间隔时间
      .sum("count");// sum或者reduce都可以
      //.reduce((a,b)=>WordWithCount(a.word,a.count+b.count))

    //打印到控制台
    windowCounts.print().setParallelism(1);

    //执行任务
```

```
            env.execute("Socket window count");

    }

    case class WordWithCount(word: String,count: Long)

}
```

2.3.3　执行程序

在前面的案例代码中指定hostname为hadoop100，port默认为9000，表示流处理程序默认监听这个主机的9000端口。因此在执行程序之前，需要先在hadoop100这个节点上面监听这个端口，通过执行下面命令实现。

```
[root@hadoop100 soft]# nc -l 9000
a
b
a
```

然后在IDEA中运行编写完成的程序代码，结果如下。

```
WordWithCount{word='a', count=1}
WordWithCount{word='b', count=1}
WordWithCount{word='a', count=2}
WordWithCount{word='b', count=1}
WordWithCount{word='a', count=1}
```

2.4　Flink批处理（Batch）案例开发

前面使用Flink实现了一个典型的流式计算案例，下面来看一下Flink的另一个应用场景——Batch离线批处理。

2.4.1　Java代码开发

需求：统计一个文件中的单词出现的总次数，并且把结果存储到文件中。

Java代码实现如下。

```java
package xuwei.tech.batch;

import org.apache.Flink.api.common.functions.FlatMapFunction;
import org.apache.Flink.api.Java.DataSet;
import org.apache.Flink.api.Java.ExecutionEnvironment;
import org.apache.Flink.api.Java.operators.DataSource;
import org.apache.Flink.api.Java.tuple.Tuple2;
import org.apache.Flink.util.Collector;

/**
 * 单词计数之离线计算
 *
 * Created by xuwei.tech
 */
public class BatchWordCountJava {

    public static void main(String[] args) throws Exception{
        String inputPath = "D:\\data\\file";
        String outPath = "D:\\data\\result";

        //获取运行环境
        ExecutionEnvironment env = ExecutionEnvironment.getExecutionEnvironment();
        //获取文件中的内容
        DataSource<String> text = env.readTextFile(inputPath);

        DataSet<Tuple2<String, Integer>> counts = text.flatMap(new Tokenizer()).groupBy(0).sum(1);
        counts.writeAsCsv(outPath,"\n"," ").setParallelism(1);
        env.execute("batch word count");

    }


    public static class Tokenizer implements FlatMapFunction<String,Tuple2<String, Integer>>{
        public void flatMap(String value, Collector<Tuple2<String, Integer>> out) throws Exception {
            String[] tokens = value.toLowerCase().split("\\W+");
            for (String token: tokens) {
                if(token.length()>0){
```

```
                                out.collect(new Tuple2<String, Integer>(token,1));
                    }
                }
            }
        }
    }
```

2.4.2 Scala代码开发

需求：统计一个文件中的单词出现的总次数，并且把结果存储到文件中。

Scala代码实现如下。

```
package xuwei.tech.batch

import org.apache.Flink.api.Scala.ExecutionEnvironment

/**
 * 单词计数之离线计算
 * Created by xuwei.tech
 */
object BatchWordCountScala {

  def main(args: Array[String]): Unit = {
    val inputPath = "D:\\data\\file"
    val outPut = "D:\\data\\result"

    val env = ExecutionEnvironment.getExecutionEnvironment
    val text = env.readTextFile(inputPath)

    //引入隐式转换
    import org.apache.Flink.api.Scala._

    val counts = text.flatMap(_.toLowerCase.split("\\W+"))
      .filter(_.nonEmpty)
      .map((_,1))
      .groupBy(0)
      .sum(1)
    counts.writeAsCsv(outPut,"\n"," ").setParallelism(1)
    env.execute("batch word count")
  }

}
```

2.4.3 执行程序

首先,代码中指定的 inputPath 是 D:\\data\\file 目录,我们需要在这个目录下面创建一些文件,并在文件中输入一些单词。

```
D:\data\file>dir
2018/03/20  09:01          24 a.txt
D:\data\file>type a.txt
hello a hello b
hello a
```

然后,在 IDEA 中运行程序代码,产生的结果会被存储到 outPut 指定的 D:\\data\\result 文件中。

```
D:\data>type result
hello 3
b 1
a 2
```

第3章
Flink 的安装和部署

我们对 Flink 有了一个基本的认识，并且也掌握了 Flink 程序的开发步骤。下面就来看一下如何安装和部署一个 Flink 集群，并在集群上真正运行 Flink 程序。

Flink 的安装和部署主要分为本地模式和集群模式，其中本地模式只需直接解压就可以使用，不以修改任何参数，一般在做一些简单测试的时候使用。集群模式包含 Standalone、Flink on Yarn 等模式，适合在生产环境下面使用，且需要修改对应的配置参数。

3.1 Flink 本地模式

本地模式是安装 Flink 服务较快速的一种方式。安装 Flink 需要以下基础环境。

- Linux 环境，在这里建议使用 CentOS 6.x 的版本。
- JDK 1.8，需要配置好 JAVA_HOME 环境变量。

注意：因为 Flink 依赖 Hadoop，所以在选择 Flink 安装包的时候需要参考已有的 Hadoop 版本。我们使用的是 Hadoop 2.7.5 版本。因为在这里我们使用 Flink 1.6.1 版本，所以下载的 Flink 对应版本就是 flink-1.6.1-bin-hadoop27-scala_2.11.tgz。

要将下载的 Flink 安装包上传到 Linux 机器的 /data/soft 目录下，如果 /data/soft 目录不存在则需要提前创建。

注意：在这里使用的Linux机器IP为192.168.1.100，hostname为hadoop100。

Flink本地模式的具体安装步骤如下。

（1）在Linux的Shell命令行中进入/data/soft目录，执行解压命令。

```
[root@hadoop100 soft]# tar -zxvf flink-1.6.1-bin-hadoop27-scala_2.11.tgz
flink-1.6.1/
flink-1.6.1/bin/
flink-1.6.1/bin/pyFlink.sh
flink-1.6.1/bin/config.sh
flink-1.6.1/bin/mesos-taskmanager.sh
...
```

（2）执行启动命令。

```
[root@hadoop100 soft]# cd flink-1.6.1
[root@hadoop100 flink-1.6.1]# bin/start-cluster.sh
Starting cluster.
Starting standalonesession daemon on host hadoop100.
Starting taskexecutor daemon on host hadoop100.
```

启动成功的效果如图3.1所示。

```
[root@hadoop100 flink-1.6.1]# bin/start-cluster.sh
Starting cluster.
Starting standalonesession daemon on host hadoop100.
Starting taskexecutor daemon on host hadoop100.
[root@hadoop100 flink-1.6.1]# jps
14535 Jps
13930 StandaloneSessionClusterEntrypoint
14363 TaskManagerRunner
[root@hadoop100 flink-1.6.1]#
```

图3.1 Flink启动成功界面

（3）启动成功以后，可以通过浏览器访问Flink的Web界面。

此时如果通过主机名进行访问，需要保证Windows的hosts文件对hadoop100这台机器的主机名和IP的映射关系进行了配置。

正常访问的页面效果如图3.2所示。

注意：如果想关闭Flink，则需要执行bin/stop-cluster.sh。

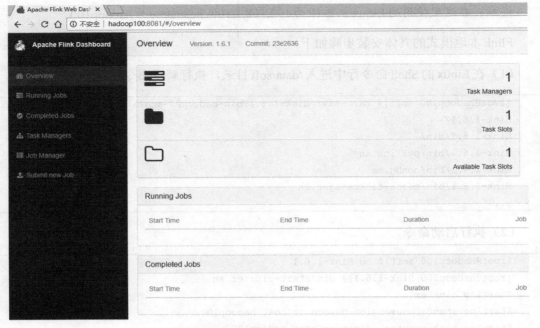

图3.2 Flink Web UI界面

3.2 Flink集群模式

Flink提供了多种集群模式,如下所示。

- Standalone。
- Flink on Yarn。
- Mesos。
- Docker。
- Kubernetes。
- AWS。
- Goole Compute Engine。
- MapR。

目前在企业中使用最多的是 Flink on Yarn 模式。在这里我们主要分析 Standalone 和 Flink on Yarn 两种模式。

3.2.1 Standalone 模式

Standalone 是 Flink 的独立部署模式，它不依赖其他平台。如果想搭建一套独立的 Flink 集群，可以考虑使用这种模式。

在使用这种模式搭建 Flink 集群之前，需要先规划集群机器信息。在这里为了搭建一个标准的 Flink 集群，需要准备 3 台 Linux 机器，如图 3.3 所示。

这 3 台 Linux 机器的基本环境信息如下。

- Master 节点，主机名是 hadoop100，JDK 1.8。
- Slave1 节点，主机名是 hadoop101，JDK 1.8。
- Slave2 节点，主机名是 hadoop102，JDK 1.8。

注意：这 3 台 Linux 机器（节点）都配置了 JAVA_HOME，并且也实现了节点之间的 ssh 免密码登录，至少保证 Master 节点可以免密码登录到其他两个 Slave 节点，防火墙也都关闭了。

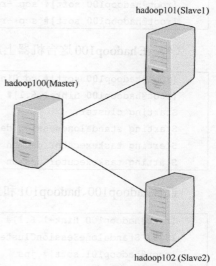

图 3.3　Flink 集群节点信息

在 Flink 集群中，Master 节点上会运行 JobManager 进程，Slave 节点上会运行 TaskManager 进程。

下面在这 3 个机器上开始正式安装和部署 Flink Standalone 模式集群。

（1）在这 3 台机器上创建 /data/soft 目录。

（2）在 hadoop100 这台机器上操作。

- 将 Flink 安装包上传到 /data/soft 目录下。
- 解压 Flink 安装包 tar -zxvf flink-1.6.1-bin-hadoop27-scala_2.11.tgz。
- 修改 Flink 的配置文件 flink-1.6.1/conf/flink-conf.yaml 中的 jobmanager.rpc.address 参数，把值修改为 hadoop100。

```
jobmanager.rpc.address: hadoop100
```

- 修改Flink的配置文件flink-1.6.1/conf/slaves，增加以下两行内容。

```
hadoop101
hadoop102
```

（3）将hadoop100这台机器上修改后的flink-1.6.1目录复制到其他两个Slave节点。

```
[root@hadoop100 soft]# scp -rq /data/soft/flink-1.6.1 hadoop101:/data/soft
[root@hadoop100 soft]# scp -rq /data/soft/flink-1.6.1 hadoop102:/data/soft
```

（4）在hadoop100这台机器上启动Flink集群服务。

```
[root@hadoop100 soft]# cd flink-1.6.1
[root@hadoop100 flink-1.6.1]# bin/start-cluster.sh
Starting cluster.
Starting standalonesession daemon on host hadoop100.
Starting taskexecutor daemon on host hadoop101.
Starting taskexecutor daemon on host hadoop102.
```

查看hadoop100、hadoop101和hadoop102这3个节点上的进程信息。

```
[root@hadoop100 flink-1.6.1]# jps
18387 StandaloneSessionClusterEntrypoint
[root@hadoop101 soft]# jps
1843 TaskManagerRunner
[root@hadoop102 soft]# jps
1944 TaskManagerRunner
```

（5）查看Flink Web UI界面，访问http://hadoop100:8081，如图3.4所示。

注意：在Flink 1.6版本中，Master节点的进程名称变成了StandaloneSessionClusterEntrypoint，Slave节点的进程名称变成了TaskManagerRunner。在这里为了方便使用，还沿用之前版本的称呼，把Master节点启动的进程称为JobManager，Slave节点的进程称为TaskManager。

如果集群中Master节点的JobManager进程停止，或者由于机器重启导致进程停止，可以通过下面的命令进行启动。

```
[root@hadoop100 flink-1.6.1]# bin/jobmanager.sh start
Starting standalonesession daemon on host hadoop100.
```

3.2 Flink 集群模式   25

图3.4　Flink Standalone集群启动成功的效果

同理，如果想要手工停止JobManager进程，则可以通过如下命令实现。

```
[root@hadoop100 flink-1.6.1]# bin/jobmanager.sh stop
Stopping standalonesession daemon (pid: 19762) on host hadoop100.
```

如果集群中Slave节点的TaskManager进程停止，或者想要向正在运行的集群中增加新的Slave节点，则可以使用下面的命令进行启动。

```
[root@hadoop101 flink-1.6.1]# bin/taskmanager.sh start
Starting taskexecutor daemon on host hadoop101.
```

同理，如果想要手工停止TaskManager进程，则可以通过如下命令实现。

```
[root@hadoop101 flink-1.6.1]# bin/taskmanager.sh stop
Stopping taskexecutor daemon (pid: 2325) on host hadoop101.
```

下面针对flink-conf.yaml文件中的几个重要参数进行分析。

- jobmanager.heap.mb：JobManager节点可用的内存大小。

- taskmanager.heap.mb：TaskManager节点可用的内存大小。

- taskmanager.numberOfTaskSlots：每台机器可用的CPU数量。

- parallelism.default：默认情况下Flink任务的并行度。
- taskmanager.tmp.dirs：TaskManager的临时数据存储目录。

上面参数中所说的Slot和parallelism的区别如下。

- Slot是静态的概念，是指TaskManager具有的并发执行能力。
- parallelism是动态的概念，是指程序运行时实际使用的并发能力。
- 设置合适的parallelism能提高运算效率。

3.2.2 Flink on Yarn模式

Flink on Yarn模式的原理是依靠YARN来调度Flink任务，目前在企业中使用较多。这种模式的好处是可以充分利用集群资源，提高集群机器的利用率，并且只需要1套Hadoop集群，就可以执行MapReduce和Spark任务，还可以执行Flink任务等，操作非常方便，不需要维护多套集群，运维方面也很轻松。Flink on Yarn模式需要依赖Hadoop集群，并且Hadoop的版本需要是2.2及以上。

Flink on Yarn模式在使用的时候又可以分为两种，如图3.5所示。

- 第1种模式，是在YARN中提前初始化一个Flink集群(称为Flink yarn-session)，开辟指定的资源，以后的Flink任务都提交到这里。这个Flink集群会常驻在YARN集群中，除非手工停止。这种方式创建的Flink集群会独占资源，不管有没有Flink任务在执行，YARN上面的其他任务都无法使用这些资源。
- 第2种模式，每次提交Flink任务都会创建一个新的Flink集群，每个Flink任务之间相互独立、互不影响，管理方便。任务执行完成之后创建的Flink集群也会消失，不会额外占用资源，按需使用，这使资源利用率达到最大，在工作中推荐使用这种模式。

下面分别介绍这两种模式的使用步骤。

1. 第1种模式

（1）创建一个一直运行的Flink集群(也可以称为Flink yarn-session)。

```
[root@hadoop100 flink-1.6.1]# bin/yarn-session.sh -n 2 -jm 1024 -tm 1024 -d
```

注意：参数最后面的-d参数是可选项，表示是否在后台独立运行。

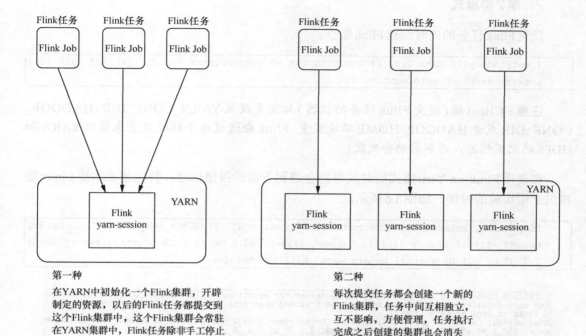

图3.5　Flink on yarn的两种模式

执行以后可以到hadoop100的8088任务界面确认是否有Flink任务成功运行。

（2）附着到一个已存在的Flink集群中。

```
[root@hadoop100 flink-1.6.1]# bin/yarn-session.sh -id applicationId
```

注意：参数最后面的applicationId是一个变量，需要获取第1步创建的Flink集群对应的applicationId信息。

（3）执行Flink任务。

```
[root@hadoop100 flink-1.6.1]# bin/flink run ./examples/batch/WordCount.jar
```

注意：WordCount.jar程序默认的输入是内置数据，默认的输出是stdout。当然也可以通过-input和-output来手动指定输入数据目录和输出数据目录。

```
-input hdfs://hadoop100:9000/LICENSE
-output hdfs://hadoop100:9000/wordcount-result.txt
```

2. 第2种模式

提交Flink任务的同时创建Flink集群。

```
[root@hadoop100 flink-1.6.1]# bin/flink run -m yarn-cluster -yn 2 -yjm 1024 -ytm 1024 ./examples/batch/WordCount.jar
```

注意：Client端（提交Flink任务的机器）必须要设置YARN_CONF_DIR、HADOOP_CONF_DIR或者HADOOP_HOME环境变量，Flink会通过这个环境变量来读取YARN和HDFS的配置信息，否则启动会失败。

在使用Flink on Yarn模式的时候可能会遇到下面的报错信息，特别是在本地Linux虚拟机上做实验的时候，如图3.6所示。

```
Diagnostics: Container [pid=6386,containerID=container_1521277661809_0006_01_000001] is running beyond virtual memory limits. Current usage: 250.5 MB of 1 GB physical memory used; 2.2 GB of 2.1 GB virtual memory used. Killing container
```

```
277661809_0006 failed 1 times due to AM Container for appattempt_1521277661809_0006_000001 exited
For more detailed output, check application tracking page:http://hadoop100:8088/cluster/app/appli
n links to logs of each attempt.
Diagnostics: Container [pid=6386,containerID=container_1521277661809_0006_01_000001] is running b
sage: 250.5 MB of 1 GB physical memory used; 2.2 GB of 2.1 GB virtual memory used. Killing contai
Dump of the process-tree for container_1521277661809_0006_01_000001 :
    |- PID PPID PGRPID SESSID CMD_NAME USER_MODE_TIME(MILLIS) SYSTEM_TIME(MILLIS) VMEM_USAGE(
NE
    |- 6386 6384 6386 6386 (bash) 0 0 108625920 331 /bin/bash -c /usr/local/jdk/bin/java -Xmx
s/userlogs/application_1521277661809_0006/container_1521277661809_0006_01_000001/jobmanager.log -
ties org.apache.flink.yarn.YarnApplicationMasterRunner 1> /usr/local/hadoop/logs/userlogs/applic
1277661809_0006_01_000001/jobmanager.out 2> /usr/local/hadoop/logs/userlogs/application_152127766
_01_000001/jobmanager.err
    |- 6401 6386 6386 6386 (java) 388 72 2287009792 63800 /usr/local/jdk/bin/java -Xmx424m -D
gs/application_1521277661809_0006/container_1521277661809_0006_01_000001/jobmanager.log -Dlog4j.c
.apache.flink.yarn.YarnApplicationMasterRunner

Container killed on request. Exit code is 143
Container exited with a non-zero exit code 143
Failing this attempt. Failing the application.
2018-03-17 21:50:21,697 INFO  org.apache.flink.yarn.ApplicationClient               - Sen
```

图3.6 Flink on Yarn常见报错信息

出现这个错误的原因是使用容量超过了虚拟内存，可以选择禁用此检查机制。

解决方法是修改文件$HADOOP_HOME/etc/hadoop/yarn-site.xml，在其中添加如下配置。

```
<property>
    <name>yarn.nodemanager.vmem-check-enabled</name>
    <value>false</value>
</property>
```

Flink on Yarn的内部实现如图3.7所示。

YARN客户端需要访问Hadoop配置,从而连接YARN资源管理器和HDFS。使用下面的策略来决定Hadoop配置。

测试是否设置了YARN_CONF_DIR、HADOOP_CONF_DIR或HADOOP_CONF_PATH环境变量(按该顺序测试)。如果设置了任意一个,就会用其来读取配置。

如果上面的策略失败了(在正确安装YARN的情况下,这不会发生),客户端就会使用HADOOP_HOME环境变量。如果已经设置了该变量,客户端就会尝试访问$HADOOP_HOME/etc/Hadoop (Hadoop 2)和$HADOOP_HOME/conf(Hadoop 1)。

当启动一个新的Flink YARN Client会话时,客户端首先会检查所请求的资源(容器和内存)是否可用。之后,它会上传Flink配置和JAR文件到HDFS。

客户端的下一步是请求一个YARN容器启动ApplicationMaster。JobManager和ApplicationMaster(AM)运行在同一个容器中,一旦它们成功地启动了,AM就能够知道JobManager的地址,它会为TaskManager生成一个新的Flink配置文件(这样它才能连上JobManager),该文件也同样会被上传到HDFS。另外,AM容器还提供了Flink的Web界面服务。Flink用来提供服务的端口是由用户和应用程序ID作为偏移配置的,这使得用户能够并行执行多个YARN会话。

之后,AM开始为Flink的TaskManager分配容器,从HDFS下载JAR文件和修改过的配置文件。一旦这些步骤完成了,Flink就安装完成并准备接受任务了。

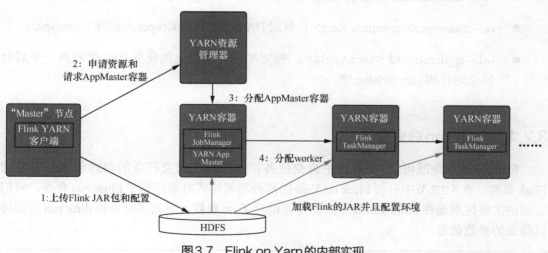

图3.7　Flink on Yarn的内部实现

3.2.3 yarn-session.sh命令分析

我们可以通过yarn-session.sh来创建Flink集群，其中yarn-session.sh后面支持多个参数。下面针对一些常见的参数进行分析。

1．必选参数

-n,--container <arg> 表示分配容器的数量（也就是TaskManager的数量）。

2．可选参数

- -D <arg> 动态属性。
- -d,--detached 在后台独立运行。
- -jm,--jobManagerMemory <arg>：设置JobManager的内存，单位是MB。
- -nm,--name：在YARN上为一个自定义的应用设置一个名字。
- -q,--query：显示YARN中可用的资源（内存、cpu核数）。
- -qu,--queue <arg>：指定YARN队列。
- -s,--slots <arg>：每个TaskManager使用的Slot数量。
- -tm,--taskManagerMemory <arg>：每个TaskManager的内存，单位是MB。
- -z,--zookeeperNamespace <arg>：针对HA模式在ZooKeeper上创建NameSpace。
- -id,--applicationId <yarnAppId>：指定YARN集群上的任务ID，附着到一个后台独立运行的yarn session中。

3.2.4 Flink run命令分析

Flink run命令既可以向Flink中提交任务，也可以在提交任务的同时创建一个新的Flink集群。在3.2.2节中分析Flink on Yarn的两种部署模式时都用到了Flink run命令，它们之间的主要区别是在Flink run命令后面是否指定了-m参数。下面详细分析flink run后面可以指定的参数信息。

```
bin/flink run [OPTIONS] <jar-file> <arguments>
```

其中[]表示是可选参数，<>表示是必填参数。

OPTIONS可以使用下面这些参数。

- -c,--class <classname>：如果没有在JAR包中指定入口类，则需要在此通过这个参数动态指定JAR包的入口类。（注意：这个参数一定要放到<jar-file>参数前面。）

- -m,--jobmanager <host:port>：指定需要连接的JobManager（主节点）地址，可以指定一个不同于配置文件中的JobManager。

- -p,--parallelism <parallelism>：动态指定任务的并行度，可以覆盖配置文件中的默认值。

具体使用场景如下。

- [root@hadoop100 flink-1.6.1]# bin/flink run ./examples/batch/WordCount.jar -input hdfs://hadoop100:9000/hello.txt -output hdfs://hadoop100:9000/result1。

默认查找当前YARN集群中已有的yarn-session的JobManager（默认到这个路径下面查看/tmp/.yarn-properties-root），然后提交任务。

- [root@hadoop100 flink-1.6.1]# bin/flink run -m hadoop100:1234 ./examples/batch/WordCount.jar -input hdfs://hadoop100:9000/hello.txt -output hdfs://hadoop100:9000/result……

显式连接指定host和port的JobManager，然后提交任务。

- [root@hadoop100 flink-1.6.1]# bin/flink run -m yarn-cluster -yn 2 ./examples/batch/WordCount.jar -input hdfs://hadoop100:9000/hello.txt -output hdfs://hadoop100:9000/result3。

在YARN中启动一个新的Flink集群，然后提交任务。

注意：yarn-sessio.sh中的参数可以用在bin/flink上，这个时候需要在yarn-session.sh的参数前面加一个y或者yarn的前缀。示例如下。

```
[root@hadoop100 flink-1.6.1]# bin/flink run -m yarn-cluster -yn 2 ./examples/batch/WordCount.jar -input hdfs://hadoop100:9000/hello.txt -output hdfs://hadoop100:9000/result4
```

3.3 Flink代码生成JAR包

在2.3节和2.4节中我们开发了Flink的Streaming流处理程序和Batch批处理程序，当时只是在IDEA中执行了程序，现在我们想要把这个程序提交到集群中真正执行。在这里以 2.3 节中的 Streaming 程序为例，在具体执行之前，需要先把代码打成 JAR 包，具体步骤如下。

（1）在 2.3 节中使用的 Maven 项目的 pom 文件中添加如下配置。

```xml
<build>
    <plugins>
        <!-- 编译插件 -->
        <plugin>
            <groupId>org.apache.maven.plugins</groupId>
            <artifactId>maven-compiler-plugin</artifactId>
            <version>3.6.0</version>
            <configuration>
                <source>1.8</source>
                <target>1.8</target>
                <encoding>UTF-8</encoding>
            </configuration>
        </plugin>
        <!-- Scala编译插件 -->
        <plugin>
            <groupId>net.alchim31.maven</groupId>
            <artifactId>scala-maven-plugin</artifactId>
            <version>3.1.6</version>
            <configuration>
                <scalaCompatVersion>2.11</scalaCompatVersion>
                <scalaVersion>2.11.12</scalaVersion>
                <encoding>UTF-8</encoding>
            </configuration>
            <executions>
                <execution>
                    <id>compile-scala</id>
                    <phase>compile</phase>
                    <goals>
                        <goal>add-source</goal>
                        <goal>compile</goal>
                    </goals>
                </execution>
                <execution>
                    <id>test-compile-scala</id>
                    <phase>test-compile</phase>
                    <goals>
                        <goal>add-source</goal>
```

```xml
                    <goal>testCompile</goal>
                </goals>
            </execution>
        </executions>
    </plugin>
    <!-- 打JAR包插件（会包含所有依赖）-->
    <plugin>
        <groupId>org.apache.maven.plugins</groupId>
        <artifactId>maven-assembly-plugin</artifactId>
        <version>2.6</version>
        <configuration>
            <descriptorRefs>
                <descriptorRef>jar-with-dependencies</descriptorRef>
            </descriptorRefs>
            <archive>
                <manifest>
                    <!-- 可以设置JAR包的入口类（可选）-->
                    <mainClass></mainClass>
                </manifest>
            </archive>
        </configuration>
        <executions>
            <execution>
                <id>make-assembly</id>
                <phase>package</phase>
                <goals>
                    <goal>single</goal>
                </goals>
            </execution>
        </executions>
    </plugin>
</plugins>
</build>
```

（2）执行编译打JAR包命令。

```
mvn clean package -DskipTests
```

命令解释如下。

- mvn：Maven中的命令。
- clean：清空target中的内容。
- package：打包。
- -D：可以动态指定参数。
- skipTests：在打包的时候不执行测试代码。

执行完上面的打包命令以后，如果看到图3.8所示的内容，则表示打JAR包命令执行成功。此时就可以在图3.9所示的目录中找到生成的JAR包，其中FlinkExample-1.0-SNAPSHOT-jar-with-dependencies.jar是包含第三方依赖的JAR包，建议使用它。

```
[INFO] --- maven-jar-plugin:2.3.2:jar (default-jar) @ FlinkExample ---
[INFO] Building jar: D:\IdeaProjects\FlinkExample\target\FlinkExample-1.0-SNAPSHOT.jar
[INFO]
[INFO] --- maven-assembly-plugin:2.6:single (make-assembly) @ FlinkExample ---
[INFO] Building jar: D:\IdeaProjects\FlinkExample\target\FlinkExample-1.0-SNAPSHOT-jar-with-dependen
cies.jar
[INFO] ------------------------------------------------------------------------
[INFO] BUILD SUCCESS
[INFO] ------------------------------------------------------------------------
[INFO] Total time: 37.515s
[INFO] Finished at: Tue Mar 19 21:19:42 CST 2019
[INFO] Final Memory: 269M/1021M
[INFO] ------------------------------------------------------------------------
D:\IdeaProjects\FlinkExample>
```

图3.8　打JAR包命令执行成功

```
D:\IdeaProjects\FlinkExample>cd target

D:\IdeaProjects\FlinkExample\target>dir
 驱动器 D 中的卷是 软件
 卷的序列号是 0004-3DCC

 D:\IdeaProjects\FlinkExample\target 的目录

2019/03/19  21:19    <DIR>          .
2019/03/19  21:19    <DIR>          ..
2019/03/19  21:19    <DIR>          archive-tmp
2019/03/19  21:19    <DIR>          classes
2019/03/19  21:19                 1 classes.-235924572.timestamp
2019/03/19  21:19        84,712,880 FlinkExample-1.0-SNAPSHOT-jar-with-dependencies.jar
2019/03/19  21:19           299,335 FlinkExample-1.0-SNAPSHOT.jar
2019/03/19  21:19    <DIR>          generated-sources
2019/03/19  21:19    <DIR>          maven-archiver
2019/03/19  21:19    <DIR>          maven-status
2019/03/19  21:19    <DIR>          test-classes
               3 个文件     85,012,216 字节
               8 个目录 216,904,687,616 可用字节

D:\IdeaProjects\FlinkExample\target>
```

图3.9　JAR包所在目录

3.4 Flink HA 的介绍和使用

3.4.1 Flink HA

默认情况下，每个 Flink 集群只有一个 JobManager，这将导致单点故障（SPOF），如果这个 JobManager 挂了，则不能提交新的任务，并且运行中的程序也会失败。使用 JobManager HA，集群可以从 JobManager 故障中恢复，从而避免单点故障。用户可以在 Standalone 或 Flink on Yarn 集群模式下配置 Flink 集群 HA（高可用性）。

Standalone 模式下，JobManager 的高可用性的基本思想是，任何时候都有一个 Master JobManager 和多个 Standby JobManager。Standby JobManager 可以在 Master JobManager 挂掉的情况下接管集群成为 Master JobManager，这样避免了单点故障，一旦某一个 Standby JobManager 接管集群，程序就可以继续运行。Standby JobManagers 和 Master JobManager 实例之间没有明确区别，每个 JobManager 都可以成为 Master 或 Standby。

另外，Flink on Yarn 的高可用性其实主要利用 YARN 的任务恢复机制实现。

3.4.2 Flink Standalone 集群的 HA 安装和配置

在这里，我们使用 3 个领导节点（JobManager 节点），其状态的变化情况如图 3.10 所示。

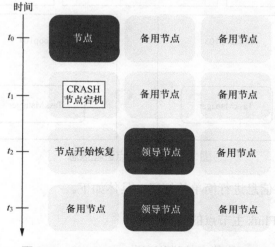

图 3.10　JobManager 节点状态的变化情况

安装和配置 Flink HA 的具体步骤如下。

1. HA集群环境规划

使用5台机器实现，其中3台 Master 节点（JobManager）和两台 Slave 节点（TaskManager）。实现 HA 还需要依赖 ZooKeeper 和 HDFS，因此要有一个 ZooKeeper 集群和 Hadoop 集群。ZooKeeper 和 JobManager 部署在相同的机器上（由于本地虚拟机个数有限，因此需要共用机器，实际生产环境中 Zookeeper 也需要单独部署在独立的服务器上）。Hadoop 集群也和 JobManager 部署在相同的机器上，集群节点进程信息如图3.11所示。

- JobManager 节点：hadoop100、hadoop101、hadoop102。
- TaskManager 节点：hadoop103、hadoop104。
- ZooKeeper 节点：hadoop100、hadoop101、hadoop102。
- Hadoop 节点：hadoop100、hadoop101、hadoop102。

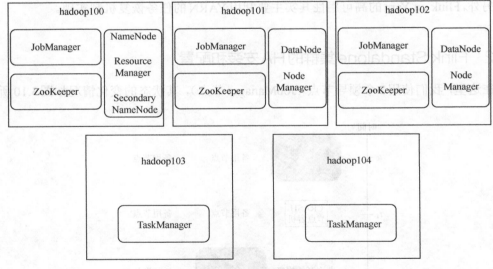

图3.11 集群节点进程信息

这里对图3.11中的信息进行简单的解释，具体如下。

- JobManager：Flink 主节点的进程名称。
- TaskManager：Flink 从节点的进程名称。

- NameNode：Hadoop中HDFS的主节点进程名称。
- DataNode：Hadoop中HDFS的从节点进程名称。
- SecondaryNameNode：Hadoop中HDFS的辅助节点名称。
- ResourceManager：Hadoop中YARN的主节点进程名称。
- NodeManager：Hadoop中YARN的从节点进程名称。
- ZooKeeper：代表ZooKeeper服务的进程。

注意：要启用JobManager高可用性，必须将高可用性模式设置为ZooKeeper，配置ZooKeeper quorum，并配置一个masters文件存储所有JobManager的hostname及其Web UI端口号。

Flink利用ZooKeeper实现运行中的JobManager节点之间的分布式协调服务。ZooKeeper是独立于Flink的服务，它通过领导选举制和轻量级状态一致性存储来提供高度可靠的分布式协调。

2．开始配置

对hadoop100、hadoop101、hadoop102、hadoop103、hadoop104这5台机器进行基础环境配置，包括设置主机名、修改/etc/hosts、设置免密码登录、关闭防火墙、设置时间同步以及安装配置JDK 1.8等内容。

注意：这里使用的机器都是CentOS 6.5版本（64位）的操作系统。

（1）设置主机名。

这5台机器的主机名分别是hadoop100、hadoop101、hadoop102、hadoop103和hadoop104。

使用hostname命令在每台机器上分别进行设置，命令如下。

在hadoop100机器上执行hostname hadoop100。

在hadoop101机器上执行hostname hadoop101。

在hadoop102机器上执行hostname hadoop102。

在hadoop103机器上执行hostname hadoop103。

在hadoop104机器上执行hostname hadoop104。

注意：使用hostname命令设置的主机名只是临时生效，因为在机器重启后，之前设置的hostname就无效了，所以还需要进行永久设置。

在hadoop100机器上执行如下命令。

```
vi /etc/sysconfig/network
HOSTNAME=hadoop100
```

在其余4台机器上执行相同的命令，仅修改HOSTNAME后的机器名称。

（2）修改/etc/hosts。

修改hadoop100、hadoop101、hadoop102、hadoop103、hadoop104这5台机器的/etc/hosts文件。

```
vi /etc/hosts
192.168.1.100 hadoop100
192.168.1.101 hadoop101
192.168.1.102 hadoop102
192.168.1.103 hadoop103
192.168.1.104 hadoop104
```

（3）设置免密码登录。

在这5台机器中，hadoop100是主节点，因此只需要实现hadoop100能够免密码登录到自己以及其他4个节点就可以了。

首先配置hadoop100节点的本机免密码登录。

```
[root@hadoop100 soft]# ssh-keygen -t rsa
[root@hadoop100 soft]# cat ~/.ssh/id_rsa.pub >> ~/.ssh/authorized_keys
```

然后在hadoop100节点上执行scp命令（执行的时候需要输入对应节点的密码）。

```
[root@hadoop100 soft]# scp ~/.ssh/authorized_keys   hadoop101:~/
[root@hadoop100 soft]# scp ~/.ssh/authorized_keys   hadoop102:~/
[root@hadoop100 soft]# scp ~/.ssh/authorized_keys   hadoop103:~/
[root@hadoop100 soft]# scp ~/.ssh/authorized_keys   hadoop104:~/
```

最后在hadoop101、hadoop102、hadoop103、hadoop104节点上分别执行如下命令。

```
cat ~/authorized_keys  >> ~/.ssh/authorized_keys
```

如果在hadoop100节点上使用ssh命令能够在不输入密码的情况下访问其他4个节点，则说明配置成功。

（4）关闭防火墙。

在hadoop100、hadoop101、hadoop102、hadoop103、hadoop104这5台机器上执行如下一行命令。

```
service iptables stop
```

注意：这样只能保证临时关闭防火墙，当机器重启以后防火墙还会自动开启。因为开启防火墙的命令在开机启动项中，所以还需要在开机启动项中关闭。执行如下一行命令。

```
chkconfig iptables off
```

（5）设置时间同步。

如果集群内节点时间相差太大的话，会导致集群服务异常，所以需要保证集群内各节点时间一致。

使用如下命令使各节点时间一致（需要保证集群内的所有节点都能够连接外网）。

```
ntpdate -u ntp.sjtu.edu.cn
```

注意：默认是没有ntpdate命令的，需要使用yum在线安装。

执行命令yum install -y ntpdate即可实现yum在线安装。

建议把这个同步时间的操作写到Linux的crontab定时器中（集群内所有节点都需要添加）。

```
vi /etc/crontab
* * * * * root /usr/sbin/ntpdate -u ntp.sjtu.edu.cn
```

（6）安装配置JDK 1.8。

① 下载JDK 1.8的安装包。

② 上传JDK 1.8的安装包。

把jdk-8u191-linux-x64.tar.gz上传到hadoop100的/data/soft目录下。如果/data/soft目录不存在，则需要创建它。

③ 解压。

```
[root@hadoop100 soft]# tar -zxvf jdk-8u191-linux-x64.tar.gz
```

④ 对 JDK 目录进行重命名。

```
[root@hadoop100 soft]# mv jdk1.8.0_191 jdk1.8
```

⑤ 配置 JAVA_HOME。

```
[root@hadoop100 soft]# vi /etc/profile
export JAVA_HOME=/data/soft/jdk1.8
export PATH=.:$JAVA_HOME/bin:$PATH
```

⑥ 验证。

```
# 首先执行source命令
[root@hadoop100 soft]# source /etc/profile
# 查看Java版本信息
[root@hadoop100 soft]# java -version
# 显示如下内容表示JDK安装配置成功
java version "1.8.0_191"
Java(TM) SE Runtime Environment (build 1.8.0_191-b12)
Java HotSpot(TM) 64-Bit Server VM (build 25.191-b12, mixed mode)
```

⑦ 把解压好的 JDK 安装目录和 /etc/profile 文件复制到其他 4 个节点上。

```
# 在hadoop100这个机器上执行scp命令
[root@hadoop100 soft]# scp -rq /data/soft/jdk1.8 hadoop101:/data/soft
[root@hadoop100 soft]# scp -rq /data/soft/jdk1.8 hadoop102:/data/soft
[root@hadoop100 soft]# scp -rq /data/soft/jdk1.8 hadoop103:/data/soft
[root@hadoop100 soft]# scp -rq /data/soft/jdk1.8 hadoop104:/data/soft
[root@hadoop100 soft]# scp -rq /etc/profile hadoop101:/etc
[root@hadoop100 soft]# scp -rq /etc/profile hadoop102:/etc
[root@hadoop100 soft]# scp -rq /etc/profile hadoop103:/etc
[root@hadoop100 soft]# scp -rq /etc/profile hadoop104:/etc
```

⑧ 在 hadoop101、hadoop102、hadoop103、hadoop104 中执行 source 命令并验证 JDK 安装和配置的情况。

```
[root@hadoop100 soft]# ssh hadoop101
[root@hadoop101 soft]# source /etc/profile
[root@hadoop101 soft]#java -version
[root@hadoop100 soft]# ssh hadoop102
[root@hadoop102 soft]# source /etc/profile
[root@hadoop102 soft]# java -version
```

```
[root@hadoop100 soft]# ssh hadoop103
[root@hadoop103 soft]# source /etc/profile
[root@hadoop103 soft]# java -version
[root@hadoop100 soft]# ssh hadoop104
[root@hadoop104 soft]# source /etc/profile
[root@hadoop104 soft]# java -version
```

（7）在hadoop100、hadoop101、hadoop102这3台机器上配置Hadoop集群。

① 下载Hadoop安装包。

```
https://archive.apache.org/dist/hadoop/common/hadoop-2.7.5/hadoop-2.7.5.tar.gz
```

② 上传Hadoop安装包。

把hadoop-2.7.5.tar.gz安装包上传到hadoop100节点的/data/soft目录下面。

③ 解压。

```
[root@hadoop100 soft]# tar -zxvf hadoop-2.7.5.tar.gz
```

④ 修改Hadoop的相关配置文件。

```
[root@hadoop100 soft]# cd hadoop-2.7.5/etc/Hadoop
[root@hadoop100 hadoop]# vi hadoop-env.sh
export JAVA_HOME=/data/soft/jdk1.8
export HADOOP_LOG_DIR=/data/hadoop_repo/logs/Hadoop

[root@hadoop100 hadoop]# vi yarn-env.sh
export JAVA_HOME=/data/soft/jdk1.8
export YARN_LOG_DIR=/data/hadoop_repo/logs/yarn

[root@hadoop100 hadoop]# vi core-site.xml
<configuration>
    <property>
        <name>fs.defaultFS</name>
        <value>hdfs://hadoop100:9000</value>
    </property>
    <property>
        <name>hadoop.tmp.dir</name>
        <value>/data/hadoop_repo</value>
    </property>
</configuration>
```

```
[root@hadoop100 hadoop]# vi hdfs-site.xml
<configuration>
    <property>
        <name>dfs.replication</name>
        <value>2</value>
    </property>
    <property>
        <name>dfs.namenode.secondary.http-address</name>
        <value>hadoop100:50090</value>
    </property>
</configuration>

[root@hadoop100 hadoop]# vi yarn-site.xml
<configuration>
    <property>
        <name>yarn.nodemanager.aux-services</name>
        <value>mapreduce_shuffle</value>
    </property>
    <property>
        <name>yarn.resourcemanager.hostname</name>
        <value>hadoop100</value>
    </property>
</configuration>

[root@hadoop100 hadoop]# mv mapred-site.xml.template  mapred-site.xml
[root@hadoop100 hadoop]# vi mapred-site.xml
<configuration>
    <property>
        <name>mapreduce.framework.name</name>
        <value>yarn</value>
    </property>
</configuration>

[root@hadoop100 hadoop]# vi slaves
hadoop101
hadoop102
```

⑤ 把hadoop100节点上修改之后的Hadoop安装目录复制到其他两个从节点。

```
# 在hadoop100节点上执行
[root@hadoop100 hadoop]# scp -rq /data/soft/hadoop-2.7.5 hadoop101:/data/soft/
[root@hadoop100 hadoop]# scp -rq /data/soft/hadoop-2.7.5 hadoop102:/data/soft/
```

⑥ 在hadoop100节点上对HDFS进行格式化。

```
[root@hadoop100 hadoop]# cd /data/soft/hadoop-2.7.5
[root@hadoop100 hadoop-2.7.5]# bin/hdfs namenode -format
```

⑦ 启动Hadoop集群。

```
# 在hadoop100节点上执行
[root@hadoop100 hadoop-2.7.5]# sbin/start-all.sh
```

⑧ 验证。

在hadoop100上执行jps命令会看到以下进程信息。

```
[root@hadoop100 hadoop-2.7.5]# jps
30101 SecondaryNameNode
29878 NameNode
30269 ResourceManager
```

在hadoop101和hadoop102上执行jps命令会看到以下进程信息。

```
[root@hadoop101 soft]# jps
4934 NodeManager
4825 DataNode
[root@hadoop102 soft]# jps
3904 NodeManager
3795 DataNode
```

⑨ 配置HADOOP_HOME环境变量。

为了后面操作Hadoop更加方便，建议配置HADOOP_HOME环境变量。

首先在hadoop100这个节点上操作，增加HADOOP_HOME变量，最终效果如下。

```
[root@hadoop100 hadoop-2.7.5]# vi /etc/profile
export JAVA_HOME=/data/soft/jdk1.8
export HADOOP_HOME=/data/soft/hadoop-2.7.5
export PATH=.:$JAVA_HOME/bin:$HADOOP_HOME/bin:$PATH
```

(8) 在hadoop100、hadoop101、hadoop102这3台机器上配置ZooKeeper集群。

① 下载ZooKeeper安装包。

```
https://archive.apache.org/dist/zookeeper/zookeeper-3.4.9/zookeeper-3.4.9.tar.gz
```

② 上传ZooKeeper安装包。

把zookeeper-3.4.9.tar.gz安装包上传到hadoop100节点的/data/soft目录下面。

③ 解压。

```
[root@hadoop100 ~]# cd /data/soft
[root@hadoop100 soft]# tar -zxvf zookeeper-3.4.9.tar.gz
```

④ 修改ZooKeeper配置文件。

```
# 进入配置文件目录
[root@hadoop100 soft]# cd zookeeper-3.4.9/conf
# 修改配置文件名称
[root@hadoop100 conf]# mv zoo_sample.cfg  zoo.cfg
# 使用vi修改文件
[root@hadoop100 conf]# vi zoo.cfg
dataDir=/data/soft/zookeeper-3.4.9/data
server.0=hadoop100:2888:3888
server.1=hadoop101:2888:3888
```

```
server.2=hadoop102:2888:3888
# 创建data目录
[root@hadoop100 conf]# cd /data/soft/zookeeper-3.4.9
[root@hadoop100 zookeeper-3.4.9]# mkdir data
# 新建文件myid，并且在文件内输入数字0
[root@hadoop100 zookeeper-3.4.9]# cd data
[root@hadoop100 data]# vi myid
0
```

⑤ 把修改好配置的ZooKeeper安装包复制到其他两个节点。

```
# 在hadoop100这个节点上执行scp命令
[root@hadoop100 data]# scp -rq /data/soft/zookeeper-3.4.9 hadoop101:/data/soft
[root@hadoop100 data]# scp -rq /data/soft/zookeeper-3.4.9 hadoop102:/data/soft
```

⑥ 修改hadoop101和hadoop102上的ZooKeeper配置。

```
# 首先修改hadoop101节点上的myid文件
[root@hadoop101 ~]# cd /data/soft/zookeeper-3.4.9/data
[root@hadoop101 data]# vi myid
1
# 然后修改hadoop102节点上的myid文件
[root@hadoop102 ~]# cd /data/soft/zookeeper-3.4.9/data
[root@hadoop102 data]# vi myid
2
```

⑦ 启动ZooKeeper集群。

```
# 分别在hadoop100、hadoop101、hadoop102上启动ZooKeeper进程
# 在hadoop100上启动
[root@hadoop100 zookeeper-3.4.9]# bin/zkServer.sh start

# 在hadoop101上启动
[root@hadoop101 zookeeper-3.4.9]# bin/zkServer.sh start

# 在hadoop102上启动
[root@hadoop102 zookeeper-3.4.9]# bin/zkServer.sh start
```

⑧ 验证。

分别在hadoop100、hadoop101、hadoop102上执行jps命令验证是否有QuorumPeerMain进程。

如果有，就说明集群启动正常了；如果没有，就到对应的节点查看zookeeper.out日志文件。

执行bin/zkServer.sh status命令会发现，有一个节点显示为leader，其他两个节点显示为follower。

（9）在hadoop100、hadoop101、hadoop102、hadoop103和hadoop104这5台机器上配置Flink，因为Flink集群内的所有节点的配置都一样，所以先从第一台机器hadoop100开始配置。

① 把Flink的安装包上传到hadoop100这个机器的/data/soft目录下。

② 解压Flink安装包。

```
[root@hadoop100 ~]# cd /data/soft
[root@hadoop100 soft]# tar -zxvf flink-1.6.1-bin-hadoop27-scala_2.11.tgz
```

③ 修改配置文件。

```
# 修改flink-conf.yaml配置文件
[root@hadoop100 soft]# cd flink-1.6.1
[root@hadoop100 flink-1.6.1]# vi conf/flink-conf.yaml
jobmanager.rpc.address: hadoop100


# 修改slaves文件
[root@hadoop100 flink-1.6.1]# vi conf/slaves
hadoop103
hadoop104


# 修改配置HA需要的参数
[root@hadoop100 flink-1.6.1]# vi conf/masters
hadoop100:8081
hadoop101:8081
hadoop102:8081


[root@hadoop100 flink-1.6.1]# vi conf/flink-conf.yaml
#要启用高可用，需要将配置文件conf/flink-conf.yaml中的high-availability mode设置修改为
zookeeper
high-availability: zookeeper
#Zookeeper的主机名和端口信息，多个参数之间用逗号隔开
high-availability.zookeeper.quorum: hadoop100:2181,hadoop101:2181,hadoop102:2181
# ZooKeeper节点根目录，其中放置所有集群节点的namespace
high-availability.zookeeper.path.root: /flink
# ZooKeeper节点集群id，其中放置了集群所需的所有协调数据
high-availability.cluster-id: /cluster_one
# 建议指定HDFS的全路径。如果某个Flink节点没有配置HDFS的话，不指定HDFS的全路径则无法识别，
storageDir存储了恢复一个JobManager所需的所有元数据。
high-availability.storageDir: hdfs://hadoop100:9000/flink/ha
```

④ 把hadoop100节点上修改配置后的flink-1.6.1目录复制到其他节点。

```
[root@hadoop100 flink-1.6.1]# scp -rq /data/soft/flink-1.6.1 hadoop101:/data/soft
[root@hadoop100 flink-1.6.1]# scp -rq /data/soft/flink-1.6.1 hadoop102:/data/soft
[root@hadoop100 flink-1.6.1]# scp -rq /data/soft/flink-1.6.1 hadoop103:/data/soft
[root@hadoop100 flink-1.6.1]# scp -rq /data/soft/flink-1.6.1 hadoop104:/data/soft
```

⑤ 确认Hadoop伪分布集群已经启动。如果没有启动，则需要先启动。

```
[root@hadoop100 flink-1.6.1]# cd /data/soft/hadoop-2.7.5
[root@hadoop100 hadoop-2.7.5]# sbin/start-all.sh
```

⑥ 确认ZooKeeper集群已经启动。如果没有启动，则需要先启动。

```
[root@hadoop100 zookeeper-3.4.9]# bin/zkServer.sh start
[root@hadoop101 zookeeper-3.4.9]# bin/zkServer.sh start
[root@hadoop102 zookeeper-3.4.9]# bin/zkServer.sh start
```

⑦ 启动Flink Standalone HA集群，在hadoop100节点上执行如下命令。

```
[root@hadoop100 flink-1.6.1]# bin/start-cluster.sh
# 启动之后会显示如下日志信息。
Starting HA cluster with 3 masters.
Starting standalonesession daemon on host hadoop100.
Starting standalonesession daemon on host hadoop101.
Starting standalonesession daemon on host hadoop102.
Starting taskexecutor daemon on host hadoop103.
Starting taskexecutor daemon on host hadoop104.
```

⑧ 验证HA集群。

查看机器进程会发现如下情况（此处只列出Flink自身的进程信息，不包含ZooKeeper和Hadoop进程信息）。

```
[root@hadoop100 flink-1.6.1]# jps
20159 StandaloneSessionClusterEntrypoint

[root@hadoop101 flink-1.6.1]# jps
7795 StandaloneSessionClusterEntrypoint

[root@hadoop102 flink-1.6.1]# jps
5046 StandaloneSessionClusterEntrypoint

[root@hadoop103 flink-1.6.1]# jps
2091 TaskManagerRunner

[root@hadoop105 flink-1.6.1]# jps
8143 TaskManagerRunner
```

因为JobManager节点都会启动Web服务，所以也可以通过Web界面进行验证。

注意：此时就算是访问hadoop101:8081，也会跳转回hadoop100:8081。因为现在hadoop100是Active的JobManager。从图3.12中也可以看出，单击Job Manager查看，显示哪个节点，就表示哪个节点现在是Active的。

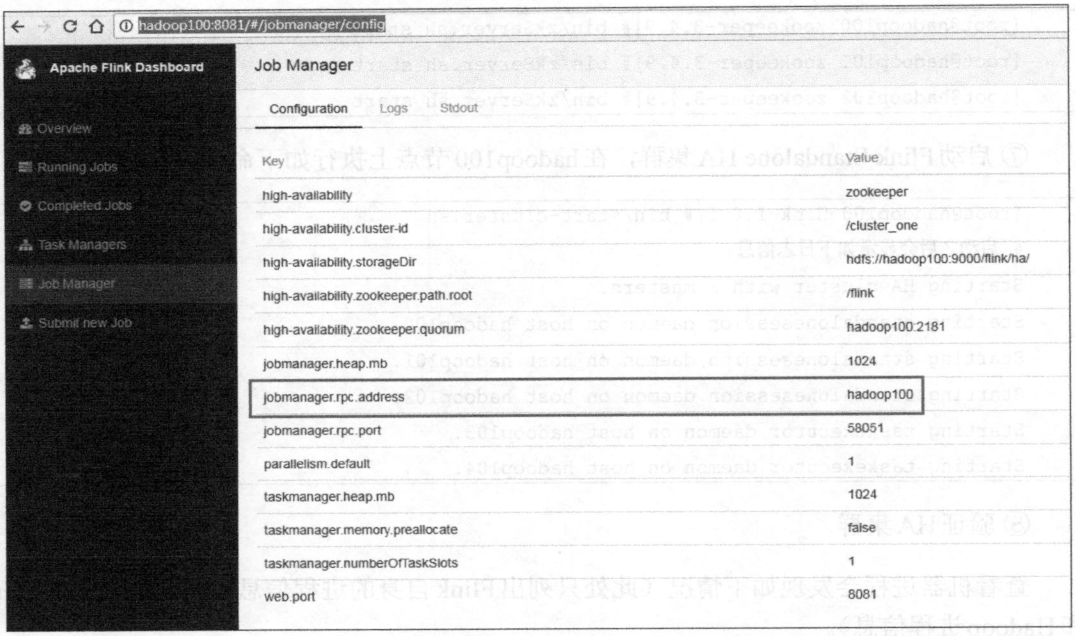

图3.12　Flink HA集群中的JobManager信息

3．模拟kill JobManager进程

现在，hadoop100节点上的JobManager是Active。这里模拟kill进程的情况，来验证hadoop101和hadoop102上的Standby状态的JobManager是否可以正常切换到Active。具体命令如下。

```
[root@hadoop100 flink-1.6.1]# jps
20159 StandaloneSessionClusterEntrypoint

# 执行kill命令
[root@hadoop100 flink-1.6.1]# kill 20159
```

4．验证JobManager切换

hadoop100节点上的JobManager进程被手工kill后，hadoop101或者hadoop102上的JobManager会自动切换为Active，中间需要时间差。

访问hadoop101上的链接http://hadoop101:8081/#/jobmanager/config，查看此链接对应页面中JobManager的信息。此时会发现Job Manager的信息变为hadoop101，如图3.13所示，则表示JobManager节点切换成功（当然此时Job Manager的信息也有可能是hadoop102）。

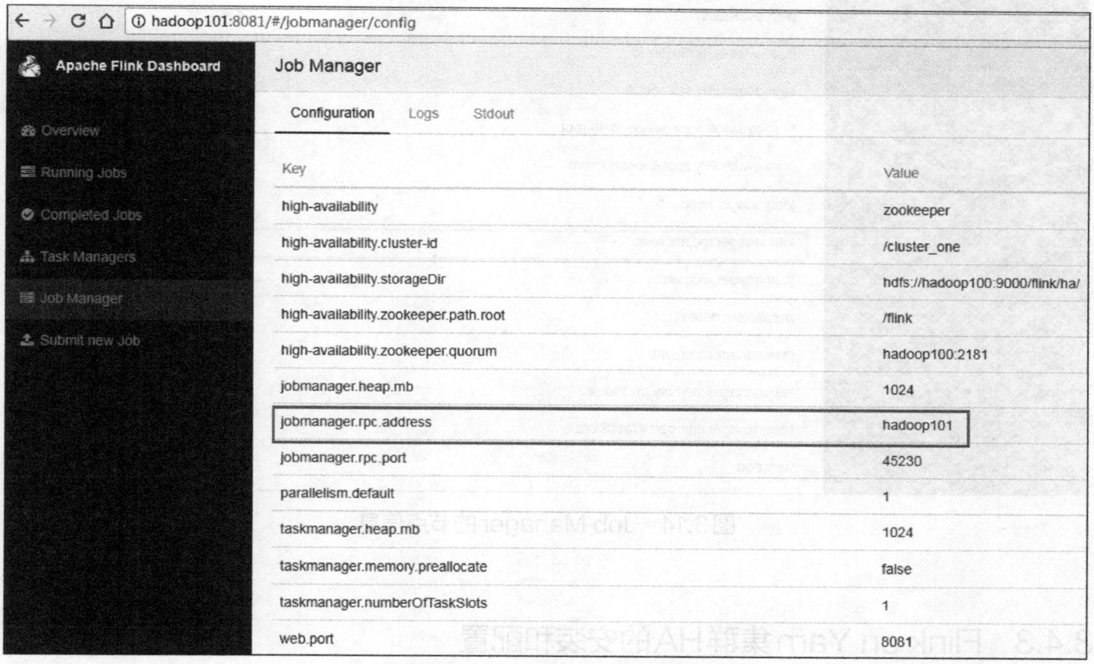

图3.13　JobManager切换后的效果

5. 重启hadoop100节点上的JobManager进程

```
# 执行如下命令启动JobManager
[root@hadoop100 flink-1.6.1]# bin/jobmanager.sh start
```

启动成功之后，可以访问如下链接查看Job Manager信息。

```
http://hadoop100:8081/#/jobmanager/config
```

这个节点重新启动之后，就变为Standby了。Job Manager的信息还是hadoop101，如图3.14所示。

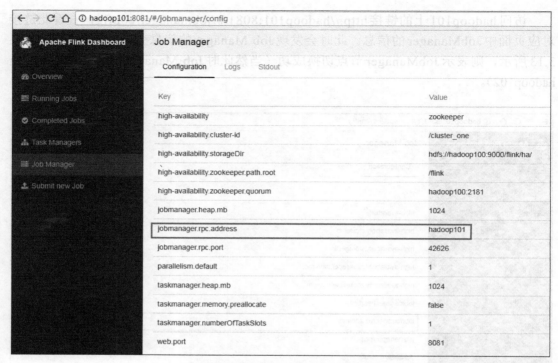

图 3.14　Job Manager 的节点信息

3.4.3　Flink on Yarn 集群 HA 的安装和配置

Flink on Yarn 的 HA 利用了 YARN 的任务恢复机制。

在这里也需要用到 ZooKeeper，主要是因为 Flink on Yarn 的 HA 虽然依赖 YARN 的任务恢复机制，但是 Flink 任务在恢复时，需要依赖检查点产生的快照。而这些快照虽然配置在 HDFS 上，但是其元数据信息保存在 ZooKeeper 中，所以我们还要配置 ZooKeeper 的信息。

Flink on Yarn 模式运行的前提是首先有一个 Hadoop 集群。Hadoop 集群可以使用 3.4.2 节中搭建的集群，ZooKeeper 集群也使用 3.4.2 节中搭建的集群。

基本的环境信息如下。

- ZooKeeper 节点：hadoop100、hadoop101、hadoop102。

- Hadoop 节点：hadoop100、hadoop101、hadoop102。

Flink on Yarn HA 配置的具体步骤如下。

1. 修改 hadoop100 中 yarn-site.xml 的配置，设置提交应用程序的最大尝试次数

```
[root@hadoop100 /]# cd /data/soft/hadoop-2.7.5/etc/Hadoop
[root@hadoop100 hadoop]# vi yarn-site.xml
<property>
  <name>yarn.resourcemanager.am.max-attempts</name>
  <value>4</value>
  <description>
    The maximum number of application master execution attempts.
  </description>
</property>

#把hadoop100节点上修改后的配置文件同步到Hadoop集群的其他节点
[root@hadoop100 hadoop]# scp -rq yarn-site.xml  hadoop101:/data/soft/hadoop-2.7.5/etc/hadoop/
[root@hadoop100 hadoop]# scp -rq yarn-site.xml hadoop102:/data/soft/hadoop-2.7.5/etc/hadoop/
```

2. 启动 Hadoop 集群

```
# 在hadoop100节点上执行启动脚本
[root@hadoop100 hadoop]# cd /data/soft/hadoop-2.7.5
[root@hadoop100 hadoop-2.7.5]# sbin/start-all.sh
```

3. 修改 Flink 相关配置

```
# 在hadoop100节点上进行操作，解压一份新的Flink安装包
[root@hadoop100 hadoop-2.7.5]# cd /data/soft/
[root@hadoop100 soft]# tar -zxvf flink-1.6.1-bin-hadoop27-scala_2.11.tgz

# 修改配置文件
[root@hadoop100 soft]# cd flink-1.6.1
[root@hadoop100 flink-1.6.1]# vi conf/flink-conf.yaml
high-availability: zookeeper
high-availability.zookeeper.quorum: hadoop100:2181,hadoop101:2181
high-availability.storageDir: hdfs://hadoop100:9000/flink/ha-yarn
high-availability.zookeeper.path.root: /flink-yarn
yarn.application-attempts: 10
```

4. 启动 Flink on Yarn 集群，测试 HA

首先确认 Hadoop 集群和 ZooKeeper 集群已经正常启动，然后在 hadoop100 节点上启动 Flink 集群。

```
[root@hadoop100 flink-1.6.1]# bin/yarn-session.sh -n 2
```

启动成功以后，到 Hadoop 的 ResourceManager 的 Web 界面上查看对应的 Flink 集群在哪个节点上，如图 3.15 所示。

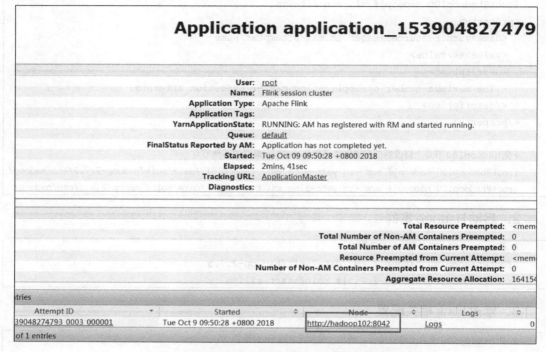

图 3.15　ResourceManager 中的任务信息

JobManager 进程就在对应节点（YarnSessionClusterEntrypoint）的进程中，想要测试 JobManager 的 HA 情况，只需要利用 YarnSessionClusterEntrypoint 进行测试即可。

执行如下命令手工模拟 kill JobManager（YarnSessionClusterEntrypoint）进程。

```
[root@hadoop100 flink-1.6.1]# ssh hadoop102
[root@hadoop102 flink-1.6.1]# jps
5325 YarnSessionClusterEntrypoint
[root@hadoop100 flink-1.6.1]# kill 5325
```

在 Web 界面进行查看，发现这个程序的 Attempt ID 变为 00002 了，说明刚才 kill 的进程重启了，如图 3.16 和图 3.17 所示。

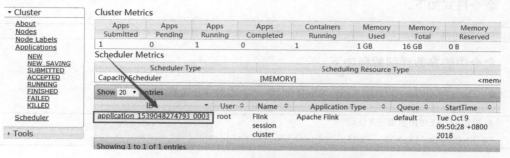

图 3.16　ResourceManager 中的任务信息

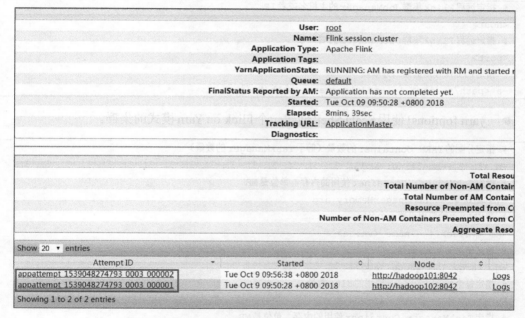

图 3.17　ResourceManager 中的 Attempt ID 信息

3.5　Flink Scala Shell

　　初学者开发的时候容易出错，每次都打包进行调试会比较麻烦，并且也不好定位问题，这时可以在 Scala Shell 命令行下进行调试。Scala Shell 方式支持流处理和批处理。当启动 Scala Shell 命令行之后，两个不同的 ExecutionEnvironments 会被自动创建。使用 senv

（Stream）和 benv（Batch）分别去处理流数据和批数据（类似于 Spark-Shell 中的 sc 变量）。

命令格式如下。

```
bin/start-scala-shell.sh [local|remote|yarn] [options] <args>
```

命令中的参数详细解释如下。

- local [options] 使用 Scala Shell 创建一个本地 Flink 集群。

```
# 指定Flink使用的第三方依赖
-a <path/to/jar> | --addclasspath <path/to/jar>
```

- remote [options] <host> <port> 使用 Scala Shell 连接一个远程 Flink 集群。

```
# 指定远程Flink集群JobManager的主机名或者IP
<host>
# 指定远程Flink集群JobManager的端口号
<port>
# 指定Flink使用的第三方依赖
-a <path/to/jar> | --addclasspath <path/to/jar>
```

- yarn [options] 使用 Scala Shell 创建一个 Flink on Yarn 模式的集群。

```
# 指定分配的YARN Container的数量（等于TaskManager的数量）
-n arg | --container arg
# 指定JobManager Container使用的内存，单位是MB
-jm arg | --jobManagerMemory arg
# 在YARN上给应用设置一个名字
-nm <value> | --name <value>
# 指定使用的YARN队列
-qu <arg> | --queue <arg>
# 指定每个TaskManager使用的Slot数量
-s <arg> | --slots <arg>
# 指定TaskManager Container使用的内存，单位是MB
-tm <arg> | --taskManagerMemory <arg>
# 指定Flink使用的第三方依赖
-a <path/to/jar> | --addClasspath <path/to/jar>
# 指定配置文件目录
--configDir <value>
# 打印帮助信息
-h | --help
```

参考案例的代码如下，效果如图3.18所示。

```
[root@hadoop100 flink-1.6.1]# bin/start-scala-shell.sh local
Starting Flink Shell:
Starting local Flink cluster (host: localhost, port: 8081).
Connecting to Flink cluster (host: localhost, port: 8081).
scala>val text = benv.fromElements("hello you","hello world")
scala>val counts = text.flatMap { _.toLowerCase.split("\\W+") }.map { (_, 1) }.groupBy(0).sum(1)
scala> counts.print()
```

注意：在使用本地模式的时候，如果本机已经启动了Standalone模式的Flink，则会报错，提示端口已被占用。此时可以使用remote模式，到Flink Web界面中查看对应的host和port信息。

```
scala> val text = benv.fromElements("hello you","hello world")
text: org.apache.flink.api.scala.DataSet[String] = org.apache.flink.api.scala.DataSet@237ee2e1

scala> val counts = text.flatMap { _.toLowerCase.split("\\W+") }.map { (_, 1) }.groupBy(0).sum(1)
counts: org.apache.flink.api.scala.AggregateDataSet[(String, Int)] = org.apache.flink.api.scala.A

scala> counts.print()
(hello,2)
(world,1)
(you,1)

scala>
```

图3.18　Flink Scala Shell执行效果

第4章
Flink常用API详解

本章主要针对Flink DataStream和DataSet的常用API进行分析和讲解，也会涉及Flink TableAPI和Flink SQL的一些常见操作。

4.1 Flink API的抽象级别分析

Flink中提供了4种不同层次的API，如图4.1所示，每种API在简洁和易用之间有自己的权衡，适用于不同的场景。目前其中的3种API用得比较多，下面自下向上介绍这4种API。

- 低级API：提供了对时间和状态的细粒度控制，简洁性和易用性较差，主要应用在对一些复杂事件的处理逻辑上。

- 核心API：主要提供了针对流数据和离线数据的处理，对低级API进行了一些封装，提供了filter、sum、max、min等高级函数，简单且易用，所以在工作中应用比较广泛。

- Table API：一般与DataSet或者DataStream紧密关联，首先通过一个DataSet或DataStream创建出一个Table；然后用类似于filter、join或者select关系型转化操作来转化为一个新的Table对象；最后将一个Table对象转回一个DataSet或DataStream。与SQL不同的是，Table API的查询不是一个指定的SQL字符串，而是调用指定的API方法。

- SQL：Flink的SQL集成是基于Apache Calcite的，Apache Calcite实现了标准的

SQL，使用起来比其他API更加灵活，因为可以直接使用SQL语句。Table API和SQL可以很容易地结合在一块使用，它们都返回Table对象。

图4.1　Flink API抽象级别

4.2　Flink DataStream的常用API

DataStream API主要分为3块：DataSource、Transformation、Sink。

- DataSource是程序的数据源输入，可以通过StreamExecutionEnvironment.addSource(sourceFunction)为程序添加一个数据源。
- Transformation是具体的操作，它对一个或多个输入数据源进行计算处理，比如Map、FlatMap和Filter等操作。
- Sink是程序的输出，它可以把Transformation处理之后的数据输出到指定的存储介质中。

4.2.1　DataSource

Flink针对DataStream提供了大量的已经实现的DataSource（数据源）接口，比如下面4种。

1．基于文件

```
readTextFile(path)
```

读取文本文件，文件遵循TextInputFormat逐行读取规则并返回。

2. 基于 Socket

```
socketTextStream
```

从 Socket 中读取数据，元素可以通过一个分隔符分开。

3. 基于集合

```
fromCollection(Collection)
```

通过 Java 的 Collection 集合创建一个数据流，集合中的所有元素必须是相同类型的。

4. 自定义输入

addSource 可以实现读取第三方数据源的数据。

Flink 也提供了一批内置的 Connector（连接器）。连接器会提供对应的 Source 支持，如表 4.1 所示。

表 4.1 Flink 内置的连接器信息

连接器	是否提供 Source 支持	是否提供 Sink 支持
Apache Kafka	是	是
Apache Cassandra	否	是
Amazon Kinesis Data Streams	是	是
Elasticsearch	否	是
Hadoop FileSystem	否	是
RabbitMQ	是	是
Apache NiFi	是	是
Twitter Streaming API	是	否

Flink 通过 Apache Bahir 组件提供了对这些连接器的支持，如表 4.2 所示。

表 4.2 Flink 通过 Apache Bahir 组件支持的连接器信息

连接器	是否提供 Source 支持	是否提供 Sink 支持
Apache ActiveMQ	是	是
Apache Flume	否	是
Redis	否	是
Akka	否	是
Netty	是	否

注意：Flink提供的这些数据源接口的容错性保证如表4.3所示。

表4.3 DataSource提供的容错情况

DataSource	语义保证	备注
File	Exactly-once（仅一次）	
Collection	Exactly-once（仅一次）	
Socket	At-most-once（最多一次）	
Kafka	Exactly-once（仅一次）	需要使用0.10及以上版本

当然也可以自定义数据源，有两种方式实现。

- 通过实现SourceFunction接口来自定义无并行度（也就是并行度只能为1）的数据源。

- 通过实现ParallelSourceFunction接口或者继承RichParallelSourceFunction来自定义有并行度的数据源。

需求：实现并行度只能为1的自定义DataSource以及SourceFunction接口。

分析：模拟产生从1开始的递增数字，每次递增加1。

Java代码实现如下。

```
package xuwei.tech.streaming.custormSource;

import org.apache.Flink.streaming.api.functions.source.SourceFunction;

/**
 * 自定义实现并行度为1的Source
 * 注意：
 * SourceFunction和SourceContext都需要指定数据类型（泛型）
 * 如果不指定，代码运行的时候会报错
 * Caused by: org.apache.Flink.api.common.functions.InvalidTypesException:
 * The types of the interface org.apache.Flink.streaming.api.functions.source.SourceFunction could not be inferred.
 * Support for synthetic interfaces, lambdas, and generic or raw types is limited at this point
 *
 *
 * Created by xuwei.tech
```

```java
    */
    public class MyNoParalleSource implements SourceFunction<Long>{

        private long count = 1L;

        private boolean isRunning = true;

        /**
         * 主要的方法
         * 启动一个Source
         * 大部分情况下，都需要在这个run方法中实现一个循环
         * 这样就可以循环产生数据了
         *
         * @param ctx
         * @throws Exception
         */
        @Override
        public void run(SourceContext<Long> ctx) throws Exception {
            while(isRunning){
                ctx.collect(count);
                count++;
                //每秒产生一条数据
                Thread.sleep(1000);
            }
        }

        /**
         * 执行cancel操作的时候会调用的方法
         *
         */
        @Override
        public void cancel() {
            isRunning = false;
        }
    }
```

Scala代码实现如下。

```
package xuwei.tech.streaming.custormSource

import org.apache.Flink.streaming.api.functions.source.SourceFunction
import org.apache.Flink.streaming.api.functions.source.SourceFunction.SourceContext
```

```scala
/**
 * 自定义实现并行度为1的source
 * Created by xuwei.tech
 */
class MyNoParallelSourceScala extends SourceFunction[Long]{

  var count = 1L
  var isRunning = true

  override def run(ctx: SourceContext[Long]) = {
    while(isRunning){
      ctx.collect(count)
      count+=1
      Thread.sleep(1000)
    }

  }

  override def cancel() = {
    isRunning = false
  }
}
```

需求：实现支持多并行度的自定义DataSource以及ParallelSourceFunction接口。

分析：模拟产生从1开始的递增数字，每次递增加1。

Java代码实现如下。

```java
package xuwei.tech.streaming.custormSource;

import org.apache.Flink.streaming.api.functions.source.ParallelSourceFunction;

/**
 * 自定义实现一个支持多并行度的Source
 * Created by xuwei.tech
 */
public class MyParalleSource implements ParallelSourceFunction<Long> {

    private long count = 1L;
```

```java
        private boolean isRunning = true;

        /**
         * 主要的方法
         * 启动一个Source
         * 大部分情况下,都需要在这个run方法中实现一个循环,这样就可以循环产生数据了
         * @param ctx
         * @throws Exception
         */
        @Override
        public void run(SourceContext<Long> ctx) throws Exception {
            while(isRunning){
                ctx.collect(count);
                count++;
                //每秒产生一条数据
                Thread.sleep(1000);
            }
        }

        /**
         * 取消一个cancel的时候会调用的方法
         */
        @Override
        public void cancel() {
            isRunning = false;
        }
    }
```

Scala代码实现如下。

```scala
package xuwei.tech.streaming.custormSource

import org.apache.Flink.streaming.api.functions.source.ParallelSourceFunction
import org.apache.Flink.streaming.api.functions.source.SourceFunction.SourceContext

/**
  * 自定义实现一个支持多并行度的Source
  * Created by xuwei.tech
```

```scala
    */
class MyParallelSourceScala extends ParallelSourceFunction[Long]{

  var count = 1L
  var isRunning = true

  override def run(ctx: SourceContext[Long]) = {
    while(isRunning){
      ctx.collect(count)
      count+=1
      Thread.sleep(1000)
    }

  }

  override def cancel() = {
    isRunning = false
  }
}
```

需求：实现支持多并行度的自定义DataSource，继承RichParallelSourceFunction类。

分析：模拟产生从1开始的递增数字，每次递增加1。

Java代码实现如下。

```java
package xuwei.tech.streaming.custormSource;

import org.apache.Flink.configuration.Configuration;
import org.apache.Flink.streaming.api.functions.source.RichParallelSourceFunction;

/**
 * 自定义实现一个支持多并行度的Source
 * RichParallelSourceFunction会额外提供open和close方法
 * 如果在source中需要获取其他链接资源，那么可以在open方法中打开资源链接，在close中关闭资源链接
 * Created by xuwei.tech
 */
public class MyRichParalleSource extends RichParallelSourceFunction<Long> {
    private long count = 1L;
```

```java
    private boolean isRunning = true;

    /**
     * 主要的方法
     * 启动一个Source
     * 大部分情况下,都需要在这个run方法中实现一个循环,这样就可以循环产生数据了
     * @param ctx
     * @throws Exception
     */
    @Override
    public void run(SourceContext<Long> ctx) throws Exception {
        while(isRunning){
            ctx.collect(count);
            count++;
            //每秒产生一条数据
            Thread.sleep(1000);
        }
    }

    /**
     * 取消一个cancel的时候会调用的方法
     */
    @Override
    public void cancel() {
        isRunning = false;
    }

    /**
     * 这个方法只会在最开始的时候被调用一次
     * 实现获取链接的代码
     * @param parameters
     * @throws Exception
     */
    @Override
    public void open(Configuration parameters) throws Exception {
        System.out.println("open......");
        super.open(parameters);
    }
```

```
    /**
     * 实现关闭链接的代码
     * @throws Exception
     */
    @Override
    public void close() throws Exception {
        super.close();
    }
}
```

Scala代码实现如下。

```
package xuwei.tech.streaming.custormSource

import org.apache.Flink.configuration.Configuration
import org.apache.Flink.streaming.api.functions.source.RichParallelSourceFunction
import org.apache.Flink.streaming.api.functions.source.SourceFunction.SourceContext

/**
  * 自定义实现一个支持多并行度的Source
  * Created by xuwei.tech
  */
class MyRichParallelSourceScala extends RichParallelSourceFunction[Long]{

  var count = 1L
  var isRunning = true

  override def run(ctx: SourceContext[Long]) = {
    while(isRunning){
      ctx.collect(count)
      count+=1
      Thread.sleep(1000)
    }

  }

  override def cancel() = {
    isRunning = false
  }

  override def open(parameters: Configuration): Unit = super.open(parameters)

  override def close(): Unit = super.close()
}
```

4.2.2 Transformation

Flink 针对 DataStream 提供了大量的已经实现的算子。

- Map：输入一个元素，然后返回一个元素，中间可以进行清洗转换等操作。
- FlatMap：输入一个元素，可以返回零个、一个或者多个元素。
- Filter：过滤函数，对传入的数据进行判断，符合条件的数据会被留下。
- KeyBy：根据指定的 Key 进行分组，Key 相同的数据会进入同一个分区。

  KeyBy 的两种典型用法如下。
  - DataStream.keyBy("someKey") 指定对象中的 someKey 段作为分组 Key。
  - DataStream.keyBy(0) 指定 Tuple 中的第一个元素作为分组 Key。

- Reduce：对数据进行聚合操作，结合当前元素和上一次 Reduce 返回的值进行聚合操作，然后返回一个新的值。
- Aggregations：sum()、min()、max() 等。
- Union：合并多个流，新的流会包含所有流中的数据，但是 Union 有一个限制，就是所有合并的流类型必须是一致的。
- Connect：和 Union 类似，但是只能连接两个流，两个流的数据类型可以不同，会对两个流中的数据应用不同的处理方法。
- coMap 和 coFlatMap：在 ConnectedStream 中需要使用这种函数，类似于 Map 和 flatMap。
- Split：根据规则把一个数据流切分为多个流。
- Select：和 Split 配合使用，选择切分后的流。

另外，Flink 针对 DataStream 提供了一些数据分区规则，具体如下。

- Random partitioning：随机分区。

```
DataStream.shuffle()
```

- Rebalancing：对数据集进行再平衡、重分区和消除数据倾斜。

```
DataStream.rebalance()
```

- Rescaling：重新调节。

```
DataStream.rescale()
```

如果上游操作有2个并发，而下游操作有4个并发，那么上游的1个并发结果分配给了下游的2个并发操作，另外的1个并发结果则分配给了下游的另外2个并发操作。另一方面，下游有2个并发操作而上游有4个并发操作，那么上游的其中2个操作的结果分配给了下游的一个并发操作，而另外2个并发操作的结果则分配给了另外1个并发操作。

Rescaling与Rebalancing的区别为Rebalancing会产生全量重分区，而Rescaling不会。

- Custom partitioning：自定义分区。

自定义分区实现Partitioner接口的方法如下。

```
DataStream.partitionCustom(partitioner, "someKey")
```

或者

```
DataStream.partitionCustom(partitioner, 0);
```

需求：创建自定义的分区规则，根据数字的奇偶性来分区。

Java代码实现如下。

```
package xuwei.tech.streaming.custormPartition;

import org.apache.Flink.api.common.functions.Partitioner;

/**
 * 自定义分区规则，根据数值的奇偶性分区
 * Created by xuwei.tech
 */
public class MyPartition implements Partitioner<Long> {
    @Override
    public int partition(Long key, int numPartitions) {
        System.out.println("分区总数:"+numPartitions);
        if(key % 2 == 0){
            return 0;
        }else{
```

```
                return 1;
            }
        }
    }
```

```java
package xuwei.tech.streaming.custormPartition;

import org.apache.Flink.api.common.functions.MapFunction;
import org.apache.Flink.api.java.tuple.Tuple1;
import org.apache.Flink.streaming.api.DataStream.DataStream;
import org.apache.Flink.streaming.api.DataStream.DataStreamSource;
import org.apache.Flink.streaming.api.environment.StreamExecutionEnvironment;
import xuwei.tech.streaming.custormSource.MyNoParalleSource;

/**
 * 使用自定义的Partition
 * Created by xuwei.tech
 */
public class SteamingDemoWithMyParitition {

    public static void main(String[] args) throws Exception{

        StreamExecutionEnvironment env = StreamExecutionEnvironment.getExecutionEnvironment();
        env.setParallelism(2);
        DataStreamSource<Long> text = env.addSource(new MyNoParalleSource());

        //对数据进行转换,把Long类型转成Tuple1类型
        DataStream<Tuple1<Long>> tupleData = text.map(new MapFunction<Long, Tuple1<Long>>() {
            @Override
            public Tuple1<Long> map(Long value) throws Exception {
                return new Tuple1<>(value);
            }
        });
        //分区之后的数据
        DataStream<Tuple1<Long>> partitionData= tupleData.partitionCustom(new MyPartition(), 0);
        DataStream<Long> result = partitionData.map(new MapFunction<Tuple1<Long>, Long>() {
```

```java
            @Override
            public Long map(Tuple1<Long> value) throws Exception {
                    System.out.println("当前线程id: " + Thread.currentThread().getId() + ",value: " + value);
                return value.getField(0);
            }
        });

        result.print().setParallelism(1);

        env.execute("SteamingDemoWithMyParitition");

    }
}
```

Scala代码实现如下。

```scala
package xuwei.tech.streaming.streamAPI

import org.apache.Flink.api.common.functions.Partitioner

/**
  * Created by xuwei.tech
  */
class MyPartitionerScala extends Partitioner[Long]{

  override def partition(key: Long, numPartitions: Int) = {
    println("分区总数: "+numPartitions)
    if(key % 2 ==0){
      0
    }else{
      1
    }


  }

}
```

```scala
package xuwei.tech.streaming.streamAPI

import java.util

import org.apache.Flink.streaming.api.collector.selector.OutputSelector
```

```scala
import org.apache.Flink.streaming.api.scala.StreamExecutionEnvironment
import xuwei.tech.streaming.custormSource.MyNoParallelSourceScala

/**
  * Created by xuwei.tech
  */
object StreamingDemoMyPartitionerScala {

  def main(args: Array[String]): Unit = {

    val env = StreamExecutionEnvironment.getExecutionEnvironment
    env.setParallelism(2)

    //隐式转换
    import org.apache.Flink.api.scala._

    val text = env.addSource(new MyNoParallelSourceScala)

    //把Long类型的数据转成Tuple类型
    val tupleData = text.map(line=>{
      Tuple1(line)// 注意Tuple1的实现方式
    })

    val partitionData = tupleData.partitionCustom(new MyPartitionerScala,0)

    val result = partitionData.map(line=>{
      println("当前线程id:"+Thread.currentThread().getId+",value: "+line)
      line._1
    })

    result.print().setParallelism(1)

    env.execute("StreamingDemoWithMyNoParallelSourceScala")
  }

}
```

4.2.3 Sink

Flink针对DataStream提供了大量的已经实现的数据目的地（Sink），具体如下所示。

- writeAsText()：将元素以字符串形式逐行写入，这些字符串通过调用每个元素的 toString() 方法来获取。

- print() / printToErr()：打印每个元素的 toString() 方法的值到标准输出或者标准错误输出流中。

- 自定义输出：addSink 可以实现把数据输出到第三方存储介质中。

系统提供了一批内置的 Connector，它们会提供对应的 Sink 支持，如表 4.1 所示。

Flink 通过 Apache Bahir 组件也提供了对这些 Connector 的支持，如表 4.2 所示。

注意：针对 Flink 提供的这些 Sink 组件，它们可以提供的容错性保证如表 4.4 所示。

表4.4　Sink 组件容错性保证

Sink	语义保证	备注
HDFS	Exactly-once	
Elasticsearch	At-least-once	
Kafka Produce	At-least-once/Exactly-once	Kafka 0.9 和 0.10 提供 At-least-once Kafka 0.11 提供 Exactly-once
File	At-least-once	
Redis	At-least-once	

当然你也可以自定义 Sink，有两种实现方式。

- 实现 SinkFunction 接口。

- 继承 RichSinkFunction 类。

自定义 Sink 代码实现建议参考 RedisSink。

```
/*
 * Licensed to the Apache Software Foundation (ASF) under one or more
 * contributor license agreements.  See the NOTICE file distributed with
 * this work for additional information regarding copyright ownership.
 * The ASF licenses this file to You under the Apache License, Version 2.0
 * (the "License"); you may not use this file except in compliance with
 * the License.  You may obtain a copy of the License at
 *
```

```
 *     http://www.apache.org/licenses/LICENSE-2.0
 * Unless required by applicable law or agreed to in writing, software
 * distributed under the License is distributed on an "AS IS" BASIS,
 * WITHOUT WARRANTIES OR CONDITIONS OF ANY KIND, either express or implied.
 * See the License for the specific language governing permissions and
 * limitations under the License.
 */

package org.apache.Flink.streaming.connectors.redis;

import org.apache.Flink.configuration.Configuration;
import org.apache.Flink.streaming.api.functions.sink.RichSinkFunction;
import org.apache.Flink.streaming.connectors.redis.common.config.FlinkJedisClusterConfig;
import org.apache.Flink.streaming.connectors.redis.common.config.FlinkJedisConfigBase;
import org.apache.Flink.streaming.connectors.redis.common.config.FlinkJedisPoolConfig;
import org.apache.Flink.streaming.connectors.redis.common.config.FlinkJedisSentinelConfig;
import org.apache.Flink.streaming.connectors.redis.common.container.RedisCommandsContainer;
import org.apache.Flink.streaming.connectors.redis.common.container.RedisCommandsContainerBuilder;
import org.apache.Flink.streaming.connectors.redis.common.mapper.RedisCommand;
import org.apache.Flink.streaming.connectors.redis.common.mapper.RedisDataType;
import org.apache.Flink.streaming.connectors.redis.common.mapper.RedisCommandDescription;
import org.apache.Flink.streaming.connectors.redis.common.mapper.RedisMapper;

import org.slf4j.Logger;
import org.slf4j.LoggerFactory;

import java.io.IOException;
import java.util.Objects;

/**
 * A sink that delivers data to a Redis channel using the Jedis client.
 * <p> The sink takes two arguments {@link FlinkJedisConfigBase} and {@link RedisMapper}.
 * <p> When {@link FlinkJedisPoolConfig} is passed as the first argument,
 * the sink will create connection using {@link redis.clients.jedis.JedisPool}. Please use this when
 * you want to connect to a single Redis server.
 * <p> When {@link FlinkJedisSentinelConfig} is passed as the first argument, the sink will create connection
```

```
 * using {@link redis.clients.jedis.JedisSentinelPool}. Please use this when you
want to connect to Sentinel.
 * <p> Please use {@link FlinkJedisClusterConfig} as the first argument if you want
to connect to
 * a Redis Cluster.
 * <p>Example:
 *
 * <pre>
 *{@code
 *public static class RedisExampleMapper implements RedisMapper<Tuple2<String, String>> {
 *    private RedisCommand redisCommand;
 *
 *    public RedisExampleMapper(RedisCommand redisCommand){
 *        this.redisCommand = redisCommand;
 *    }
 *    public RedisCommandDescription getCommandDescription() {
 *        return new RedisCommandDescription(redisCommand, REDIS_ADDITIONAL_KEY);
 *    }
 *    public String getKeyFromData(Tuple2<String, String> data) {
 *        return data.f0;
 *    }
 *    public String getValueFromData(Tuple2<String, String> data) {
 *        return data.f1;
 *    }
 *}
 *JedisPoolConfig jedisPoolConfig = new JedisPoolConfig.Builder()
 *    .setHost(REDIS_HOST).setPort(REDIS_PORT).build();
 *new RedisSink<String>(jedisPoolConfig, new RedisExampleMapper(RedisCommand.LPUSH));
 *}</pre>
 *
 * @param <IN> Type of the elements emitted by this sink
 */
public class RedisSink<IN> extends RichSinkFunction<IN> {

    private static final long serialVersionUID = 1L;

    private static final Logger LOG = LoggerFactory.getLogger(RedisSink.class);
    /**
     * This additional key needed for {@link RedisDataType#HASH} and {@link
RedisDataType#SORTED_SET}.
```

```java
     * Other {@link RedisDataType} works only with two variable i.e. name of the
list and value to be added.
     * But for {@link RedisDataType#HASH} and {@link RedisDataType#SORTED_
SET} we need three variables.
     * <p>For {@link RedisDataType#HASH} we need hash name, hash key and element.
     * {@code additionalKey} used as hash name for {@link RedisDataType#HASH}
     * <p>For {@link RedisDataType#SORTED_SET} we need set name, the element and it's
score.
     * {@code additionalKey} used as set name for {@link RedisDataType#SORTED_SET}
     */
    private String additionalKey;
    private RedisMapper<IN> redisSinkMapper;
    private RedisCommand redisCommand;

    private FlinkJedisConfigBase FlinkJedisConfigBase;
    private RedisCommandsContainer redisCommandsContainer;

    /**
     * Creates a new {@link RedisSink} that connects to the Redis server.
     *
     * @param FlinkJedisConfigBase The configuration of {@link FlinkJedisConfigBase}
     * @param redisSinkMapper This is used to generate Redis command and key value
from incoming elements.
     */
    public RedisSink(FlinkJedisConfigBase FlinkJedisConfigBase, RedisMapper<IN>
redisSinkMapper) {
        Objects.requireNonNull(FlinkJedisConfigBase, "Redis connection pool config should
not be null");
        Objects.requireNonNull(redisSinkMapper, "Redis Mapper can not be null");
        Objects.requireNonNull(redisSinkMapper.getCommandDescription(), "Redis
Mapper data type description can not be null");

        this.FlinkJedisConfigBase = FlinkJedisConfigBase;

        this.redisSinkMapper = redisSinkMapper;
        RedisCommandDescription redisCommandDescription = redisSinkMapper.
getCommandDescription();
        this.redisCommand = redisCommandDescription.getCommand();
        this.additionalKey = redisCommandDescription.getAdditionalKey();
    }
```

```java
/**
 * Called when new data arrives to the sink, and forwards it to Redis channel.
 * Depending on the specified Redis data type (see {@link RedisDataType}),
 * a different Redis command will be applied.
 * Available commands are RPUSH, LPUSH, SADD, PUBLISH, SET, PFADD, HSET, ZADD.
 * @param input The incoming data
 */
@Override
public void invoke(IN input) throws Exception {
    String key = redisSinkMapper.getKeyFromData(input);
    String value = redisSinkMapper.getValueFromData(input);

    switch (redisCommand) {
        case RPUSH:
            this.redisCommandsContainer.rpush(key, value);
            break;
        case LPUSH:
            this.redisCommandsContainer.lpush(key, value);
            break;
        case SADD:
            this.redisCommandsContainer.sadd(key, value);
            break;
        case SET:
            this.redisCommandsContainer.set(key, value);
            break;
        case PFADD:
            this.redisCommandsContainer.pfadd(key, value);
            break;
        case PUBLISH:
            this.redisCommandsContainer.publish(key, value);
            break;
        case ZADD:
            this.redisCommandsContainer.zadd(this.additionalKey, value, key);
            break;
        case ZREM:
            this.redisCommandsContainer.zrem(this.additionalKey, key);
            break;
        case HSET:
            this.redisCommandsContainer.hset(this.additionalKey, key, value);
            break;
        default:
```

```java
                    throw new IllegalArgumentException("Cannot process such data type: "
 + redisCommand);
            }
        }

        /**
         * Initializes the connection to Redis by either cluster or sentinels or single server.
         * @throws IllegalArgumentException if jedisPoolConfig, jedisClusterConfig and
jedisSentinelConfig are all null
         */
        @Override
        public void open(Configuration parameters) throws Exception {
            try {
                 this.redisCommandsContainer = RedisCommandsContainerBuilder.build(this.
FlinkJedisConfigBase);
                this.redisCommandsContainer.open();
            } catch (Exception e) {
                LOG.error("Redis has not been properly initialized: ", e);
                throw e;
            }
        }

        /**
         * Closes commands container.
         * @throws IOException if command container is unable to close.
         */
        @Override
        public void close() throws IOException {
            if (redisCommandsContainer != null) {
                redisCommandsContainer.close();
            }
        }
    }
```

下面以RedisSink为例演示具体的用法,这里面用到了Redis中的List数据类型。

需求:接收Socket传输过来的数据,把数据保存到Redis中。

注意:针对List数据类型,我们在定义getCommandDescription方法的时候,使用new RedisCommandDescription(RedisCommand.LPUSH);。

如果是Hash数据类型,在定义getCommandDescription方法的时候,需要使用new RedisCommandDescription(RedisCommand.HSET,"hashKey");，在构造函数中需要直接指定 Hash 数据类型的 Key 的名称。

Java 代码实现如下。

```java
package xuwei.tech.streaming.sink;

import org.apache.Flink.api.common.functions.MapFunction;
import org.apache.Flink.api.java.tuple.Tuple2;
import org.apache.Flink.streaming.api.DataStream.DataStream;
import org.apache.Flink.streaming.api.DataStream.DataStreamSource;
import org.apache.Flink.streaming.api.DataStream.SingleOutputStreamOperator;
import org.apache.Flink.streaming.api.environment.StreamExecutionEnvironment;
import org.apache.Flink.streaming.connectors.redis.RedisSink;
import org.apache.Flink.streaming.connectors.redis.common.config.FlinkJedisPoolConfig;
import org.apache.Flink.streaming.connectors.redis.common.mapper.RedisCommand;
import org.apache.Flink.streaming.connectors.redis.common.mapper.RedisCommandDescription;
import org.apache.Flink.streaming.connectors.redis.common.mapper.RedisMapper;

/**
 * Created by xuwei.tech
 */
public class StreamingDemoToRedis {

    public static void main(String[] args) throws Exception{
        StreamExecutionEnvironment env = StreamExecutionEnvironment.getExecutionEnvironment();

        DataStreamSource<String> text = env.socketTextStream("hadoop100", 9000, "\n");

        //lpsuh l_words word

        // 对数据进行组装,把String转化为Tuple2<String,String>
        DataStream<Tuple2<String, String>> l_wordsData = text.map(new MapFunction<String, Tuple2<String, String>>() {
            @Override
            public Tuple2<String, String> map(String value) throws Exception {
                return new Tuple2<>("l_words", value);
            }
        });
```

```java
        //创建Redis的配置
        FlinkJedisPoolConfig conf = new FlinkJedisPoolConfig.Builder().
setHost("hadoop110").setPort(6379).build();
```

```java
        //创建RedisSink
        RedisSink<Tuple2<String, String>> redisSink = new RedisSink<>(conf, new MyRedisMapper());

        l_wordsData.addSink(redisSink);

        env.execute("StreamingDemoToRedis");
    }
    public static class MyRedisMapper implements RedisMapper<Tuple2<String, String>>{
        //表示从接收的数据中获取需要操作的Redis Key
        @Override
        public String getKeyFromData(Tuple2<String, String> data) {
            return data.f0;
        }
        //表示从接收的数据中获取需要操作的Redis Value
        @Override
        public String getValueFromData(Tuple2<String, String> data) {
            return data.f1;
        }

        @Override
        public RedisCommandDescription getCommandDescription() {
            return new RedisCommandDescription(RedisCommand.LPUSH);
        }
    }
}
```

Scala代码实现如下。

```scala
package xuwei.tech.streaming.sink

import org.apache.Flink.api.java.utils.ParameterTool
import org.apache.Flink.streaming.api.scala.StreamExecutionEnvironment
import org.apache.Flink.streaming.api.windowing.time.Time
import org.apache.Flink.streaming.connectors.redis.RedisSink
import org.apache.Flink.streaming.connectors.redis.common.config.FlinkJedisPoolConfig
```

```scala
import org.apache.Flink.streaming.connectors.redis.common.mapper.{RedisCommand,
RedisCommandDescription, RedisMapper}

/**
  *
  * Created by xuwei.tech
  */
object StreamingDataToRedisScala {

  def main(args: Array[String]): Unit = {

    //获取Socket端口号
    val port = 9000

    //获取运行环境
    val env: StreamExecutionEnvironment = StreamExecutionEnvironment.getExecutionEnvironment

    //链接Socket获取输入数据
    val text = env.socketTextStream("hadoop100",port,'\n')

    //注意：必须要添加这一行隐式转行
    import org.apache.Flink.api.scala._

    val l_wordsData = text.map(line=>("l_words_scala",line))

    val conf = new FlinkJedisPoolConfig.Builder().setHost("hadoop110").setPort(6379).build()

    val redisSink = new RedisSink[Tuple2[String,String]](conf,new MyRedisMapper)

    l_wordsData.addSink(redisSink)

    //执行任务
    env.execute("Socket window count");

  }

  class MyRedisMapper extends RedisMapper[Tuple2[String,String]]{
```

```
      override def getKeyFromData(data: (String, String)) = {
        data._1
      }

      override def getValueFromData(data: (String, String)) = {
        data._2
      }

      override def getCommandDescription = {
        new RedisCommandDescription(RedisCommand.LPUSH)
      }
    }
}
```

4.3 Flink DataSet的常用API分析

DataSet API主要可以分为3块来分析：DataSource、Transformation和Sink。

- DataSource是程序的数据源输入。
- Transformation是具体的操作，它对一个或多个输入数据源进行计算处理，比如Map、FlatMap、Filter等操作。
- Sink是程序的输出，它可以把Transformation处理之后的数据输出到指定的存储介质中。

4.3.1 DataSource

对DataSet批处理而言，较频繁的操作是读取HDFS中的文件数据，因此这里主要介绍两个DataSource组件。

1．基于集合

fromCollection(Collection)，主要是为了方便测试使用。

2．基于文件

readTextFile(path)，基于HDFS中的数据进行计算分析。

4.3.2 Transformation

Flink 针对 DataSet 提供了大量的已经实现的算子。

- Map：输入一个元素，然后返回一个元素，中间可以进行清洗转换等操作。
- FlatMap：输入一个元素，可以返回零个、一个或者多个元素。
- MapPartition：类似 Map，一次处理一个分区的数据（如果在进行 Map 处理的时候需要获取第三方资源连接，建议使用 MapPartition）。
- Filter：过滤函数，对传入的数据进行判断，符合条件的数据会被留下。
- Reduce：对数据进行聚合操作，结合当前元素和上一次 Reduce 返回的值进行聚合操作，然后返回一个新的值。
- Aggregations：sum、max、min 等。
- Distinct：返回一个数据集中去重之后的元素。
- Join：内连接。
- OuterJoin：外链接。
- Cross：获取两个数据集的笛卡尔积。
- Union：返回两个数据集的总和，数据类型需要一致。
- First-n：获取集合中的前 N 个元素。
- Sort Partition：在本地对数据集的所有分区进行排序，通过 sortPartition() 的链接调用来完成对多个字段的排序。

Flink 针对 DataSet 提供了一些数据分区规则，具体如下。

- Rebalance：对数据集进行再平衡、重分区以及消除数据倾斜操作。
- Hash-Partition：根据指定 Key 的散列值对数据集进行分区。

```
partitionByHash()
```

- Range-Partition：根据指定的 Key 对数据集进行范围分区。

```
.partitionByRange()
```

- Custom Partitioning：自定义分区规则，自定义分区需要实现 Partitioner 接口。

```
partitionCustom(partitioner, "someKey")
```

或者

```
partitionCustom(partitioner, 0)
```

自定义分区的实现参考 4.2.3 节中的代码。

4.3.3 Sink

Flink 针对 DataSet 提供了大量的已经实现的 Sink。

- writeAsText()：将元素以字符串形式逐行写入，这些字符串通过调用每个元素的 toString() 方法来获取。
- writeAsCsv()：将元组以逗号分隔写入文件中，行及字段之间的分隔是可配置的，每个字段的值来自对象的 toString() 方法。
- print()：打印每个元素的 toString() 方法的值到标准输出或者标准错误输出流中。

4.4 Flink Table API 和 SQL 的分析及使用

Flink 针对标准的流处理和批处理提供了两种关系型 API：Table API 和 SQL。Table API 允许用户以一种很直观的方式进行 select、filter 和 join 操作；Flink SQL 支持基于 Apache Calcite 实现的标准 SQL。针对批处理和流处理可以提供相同的处理语义和结果。

Flink Table API、SQL 接口和 Flink 的 DataStream API、DataSet API 是紧密联系在一起的。

Table API 和 SQL 是关系型 API，用户可以像操作 MySQL 数据库表一样来操作数据，而不需要通过编写 Java 代码来完成 Flink Function，更不需要手工为 Java 代码调优。另外，SQL 作为一个非程序员可操作的语言，学习成本很低，如果一个系统提供 SQL 支持，将很容易被用户接受。

注意：Table API 和 SQL 目前正在积极开发中，有一些功能目前还是不支持的。

Flink 的 Table API 和 SQL 是捆绑在 Flink-Table 依赖中的，因此如果项目中想要使用 Table API 和 SQL，就必须要添加下面依赖。

```xml
<dependency>
    <groupId>org.apache.flink</groupId>
    <artifactId>flink-table_2.11</artifactId>
    <version>1.6.1</version>
</dependency>
```

注意：针对Flink的Scala操作，还需要添加对应的依赖。其中，针对Scala的批处理操作要添加如下依赖。

```xml
<dependency>
    <groupId>org.apache.flink</groupId>
    <artifactId>flink-scala_2.11</artifactId>
    <version>1.6.1</version>
</dependency>
```

针对Scala的流处理操作，添加如下依赖。

```xml
<dependency>
    <groupId>org.apache.flink</groupId>
    <artifactId>flink-streaming-scala_2.11</artifactId>
    <version>1.6.1</version>
</dependency>
```

注意：为了避免用户的类加载器被垃圾回收器回收，官方不建议把Flink-table这个JAR包和业务代码打包在一块，而推荐把Flink-Table的依赖复制到Flink的Lib目录下。

Table API和SQL通过join API集成在一起，这个join API的核心概念是Table，Table可以作为查询的输入和输出。下面来分析使用Table API和SQL查询程序的通用结构、如何注册Table、如何查询Table以及如何将数据发给Table。

4.4.1 Table API和SQL的基本使用

想使用Table API和SQL，首先要创建一个TableEnvironment。TableEnvironment对象是Table API和SQL集成的核心，通过TableEnvironment可以实现以下功能。

- 通过内部目录创建表。
- 通过外部目录创建表。
- 执行SQL查询。
- 注册一个用户自定义的Function。

- 把DataStream或者DataSet转换成Table。

- 持有ExecutionEnvironment或者StreamExecutionEnvironment的引用。

一个查询中只能绑定一个指定的TableEnvironment，TableEnvironment可以通过TableEnvironment.getTableEnvironment()或者TableConfig来生成。TableConfig可以用来配置TableEnvironment或者自定义查询优化。

如何创建一个TableEnvironment对象？具体实现代码如下。

Java代码实现如下。

```
//流数据查询
StreamExecutionEnvironment sEnv = StreamExecutionEnvironment.getExecutionEnvironment();
StreamTableEnvironment sTableEnv = TableEnvironment.getTableEnvironment(sEnv);

//批数据查询
ExecutionEnvironment bEnv = ExecutionEnvironment.getExecutionEnvironment();
BatchTableEnvironment bTableEnv = TableEnvironment.getTableEnvironment(bEnv);
```

Scala代码实现如下。

```
//流数据查询
val sEnv = StreamExecutionEnvironment.getExecutionEnvironment
val sTableEnv = TableEnvironment.getTableEnvironment(sEnv)
//批数据查询
val bEnv = ExecutionEnvironment.getExecutionEnvironment
val bTableEnv = TableEnvironment.getTableEnvironment(bEnv)
```

通过获取到的TableEnvironment对象可以创建Table对象，有两种类型的Table对象：输入Table(Input Table)和输出Table(Output Table)。输入Table可以给Table API和SQL提供查询数据，输出Table可以把Table API和SQL的查询结果发送到外部存储介质中。

输入Table可以通过多种数据源注册。

- 已存在的Table对象：通常是Table API和SQL的查询结果。

- TableSource：通过它可以访问外部数据，比如文件、数据库和消息队列。

- DataStream或DataSet。

输出Table需要使用TableSink注册。

下面演示如何通过TableSource注册一个Table。

Java代码实现如下。

```java
StreamExecutionEnvironment env = StreamExecutionEnvironment.getExecutionEnvironment();
StreamTableEnvironment tableEnv = TableEnvironment.getTableEnvironment(env);
//创建一个TableSource
TableSource csvSource = new CsvTableSource("/path/to/file", ...);
//注册一个TableSource,称为CvsTable
tableEnv.registerTableSource("CsvTable", csvSource);
```

Scala代码实现如下。

```scala
val env = StreamExecutionEnvironment.getExecutionEnvironment
val tableEnv = TableEnvironment.getTableEnvironment(env)
//创建一个TableSource
val csvSource: TableSource = new CsvTableSource("/path/to/file", ...)
//注册一个TableSource,称为CvsTable
tableEnv.registerTableSource("CsvTable", csvSource)
```

接下来演示如何通过TableSink把数据写到外部存储介质中。

Java代码实现如下。

```java
StreamExecutionEnvironment env = StreamExecutionEnvironment.getExecutionEnvironment();
StreamTableEnvironment tableEnv = TableEnvironment.getTableEnvironment(env);
//创建一个TableSink
TableSink csvSink = new CsvTableSink("/path/to/file", ...);
//定义字段名称和类型
String[] fieldNames = {"a", "b", "c"};
TypeInformation[] fieldTypes = {Types.INT, Types.STRING, Types.LONG};
//注册一个TableSink,称为CsvSinkTable
tableEnv.registerTableSink("CsvSinkTable", fieldNames, fieldTypes, csvSink);
```

Scala代码实现如下。

```scala
val env = StreamExecutionEnvironment.getExecutionEnvironment
val tableEnv = TableEnvironment.getTableEnvironment(env)
//创建一个TableSink
val csvSink: TableSink = new CsvTableSink("/path/to/file", ...)
```

```
//定义字段名称和类型
val fieldNames: Array[String] = Array("a", "b", "c")
val fieldTypes: Array[TypeInformation[_]] = Array(Types.INT, Types.STRING, Types.LONG)
//注册一个TableSink,称为CsvSinkTable
tableEnv.registerTableSink("CsvSinkTable", fieldNames, fieldTypes, csvSink)
```

我们知道了如何通过 TableSource 读取数据和通过 TableSink 写出数据，下面介绍如何查询 Table 中的数据。

1. 使用 Table API

Java 代码实现如下。

```
StreamExecutionEnvironment env = StreamExecutionEnvironment.getExecutionEnvironment();
StreamTableEnvironment tableEnv = TableEnvironment.getTableEnvironment(env);
//注册一个Orders表
...
//通过scan操作获取到一个Table对象
Table orders = tableEnv.scan("Orders");
//计算所有来自法国的收入
Table revenue = orders
  .filter("cCountry === 'FRANCE'")
  .groupBy("cID, cName")
  .select("cID, cName, revenue.sum AS revSum");
```

Scala 代码实现如下。

```
val env = StreamExecutionEnvironment.getExecutionEnvironment
val tableEnv = TableEnvironment.getTableEnvironment(env)
//注册一个Orders表
...
//通过scan操作获取到一个Table对象
val orders = tableEnv.scan("Orders")
//计算所有来自法国的收入
val revenue = orders
  .filter('cCountry === "FRANCE")
  .groupBy('cID, 'cName)
  .select('cID, 'cName, 'revenue.sum AS 'revSum)
```

2. 使用 SQL

Java 代码实现如下。

```
StreamExecutionEnvironment env = StreamExecutionEnvironment.getExecutionEnvironment();
StreamTableEnvironment tableEnv = TableEnvironment.getTableEnvironment(env);
//注册一个Orders表
...
//计算所有来自法国的收入
Table revenue = tableEnv.sqlQuery(
    "SELECT cID, cName, SUM(revenue) AS revSum " +
    "FROM Orders " +
    "WHERE cCountry = 'FRANCE' " +
    "GROUP BY cID, cName"
  );
```

Scala代码实现如下。

```
val env = StreamExecutionEnvironment.getExecutionEnvironment
val tableEnv = TableEnvironment.getTableEnvironment(env)
//注册一个Orders表
...
//计算所有来自法国的收入
val revenue = tableEnv.sqlQuery("""
  |SELECT cID, cName, SUM(revenue) AS revSum
  |FROM Orders
  |WHERE cCountry = 'FRANCE'
  |GROUP BY cID, cName
  """.stripMargin)
```

注意：Table API和SQL查询很容易融合在一起，因为它们都返回Table对象。

- Table API查询可以基于SQL查询结果的Table来进行。

- SQL查询可以基于Table API查询的结果来定义。

4.4.2 DataStream、DataSet和Table之间的转换

Table API和SQL查询可以很容易地和DataStream、DataSet程序集成到一起。通过一个TableEnvironment，可以把DataStream或者DataSet注册为Table，这样就可以使用Table API和SQL查询了。通过TableEnvironment也可以把Table对象转换为DataStream或者DataSet，这样就能够使用DataStream或者DataSet中的相关API了。

1. 把DataStream或者DataSet注册为Table对象

（1）通过注册的形式实现。

Java 代码实现如下。

```
StreamExecutionEnvironment env = StreamExecutionEnvironment.getExecutionEnvironment();
StreamTableEnvironment tableEnv = TableEnvironment.getTableEnvironment(env);
//获取 DataStream
DataStream<Tuple2<Long, String>> stream = ...
//把 DataStream 注册为 Table，称为 myTable，表中的字段为 f0,f1
tableEnv.registerDataStream("myTable", stream);
//在注册 Table 的时候也可以手工指定字段的名称
tableEnv.registerDataStream("myTable2", stream, "myLong, myString");
```

Scala 代码实现如下。

```
val env = StreamExecutionEnvironment.getExecutionEnvironment
val tableEnv = TableEnvironment.getTableEnvironment(env)
//获取 DataStream
val stream: DataStream[(Long, String)] = ...
//把 DataStream 注册为 Table，称为 myTable，表中的字段为 f0,f1
tableEnv.registerDataStream("myTable", stream)

//在注册 Table 的时候也可以手工指定字段的名称
import org.apache.flink.table.api.scala._
tableEnv.registerDataStream("myTable2", stream, 'myLong, 'myString)
```

注意：DataStream 程序的表名不能满足规则 ^_DataStreamTable_[0-9]+，DataSet 程序的表名不能满足规则 ^_DataSetTable_[0-9]+，这些规则的名字是内部使用的。

（2）通过直接转化的形式实现。

Java 代码实现如下。

```
StreamExecutionEnvironment env = StreamExecutionEnvironment.getExecutionEnvironment();
StreamTableEnvironment tableEnv = TableEnvironment.getTableEnvironment(env);
//获取 DataStream
DataStream<Tuple2<Long, String>> stream = ...
//把 DataStream 转化为 Table，使用默认的字段名称 f0,f1
Table table1 = tableEnv.fromDataStream(stream);
//把 DataStream 转化为 Table，使用指定的字段名称 "myLong", "myString"
Table table2 = tableEnv.fromDataStream(stream, "myLong, myString");
```

Scala 代码实现如下。

```
val env = StreamExecutionEnvironment.getExecutionEnvironment
val tableEnv = TableEnvironment.getTableEnvironment(env)
//获取DataStream
val stream: DataStream[(Long, String)] = ...
//把DataStream转化为Table，使用默认的字段名_1, '_2
val table1: Table = tableEnv.fromDataStream(stream)

//把DataStream转化为Table，使用指定的字段名称'myLong, 'myString
import org.apache.flink.table.api.scala._
val table2: Table = tableEnv.fromDataStream(stream, 'myLong, 'myString)
```

2. 把Table对象转换为DataStream或者DataSet

当我们想把一个Table对象转换为DataStream或者DataSet的时候，需要指定DataStream或者DataSet中数据的类型。通常较方便的转换类型是行，如下都是支持的数据类型。

- Row：通过角标映射字段，支持任意数量的字段，并且支持null值和非类型安全的访问。

- POJO：Java中的实体类，这个实体类中的字段名称需要和Table中的字段名称保持一致，支持任意数量的字段，支持null值，类型安全的访问。

- Case Class：通过角标映射字段，不支持null值，类型安全的访问

- Tuple：通过角标映射字段，Scala中限制22个字段，Java中限制25个字段，不支持null值，类型安全的访问。

- Atomic Type：Table必须要有一个字段，不支持null值，类型安全的访问。

（1）把Table转化为DataStream。

流式查询的结果Table会被动态地更新，即每个新的记录到达输入流时结果就会发生变化。因此，转换此动态查询的DataStream需要对表的更新进行编码。

有几种模式可以将Table转换为DataStream。

- Append Mode：这种模式只适用于当动态表仅由INSERT更改修改时（仅附加），之前添加的数据不会被更新。

- Retract Mode：可以始终使用此模式，它使用一个Boolean标识来编码INSERT和DELETE更改。

Java代码实现如下。

```java
StreamExecutionEnvironment env = StreamExecutionEnvironment.getExecutionEnvironment();
StreamTableEnvironment tableEnv = TableEnvironment.getTableEnvironment(env);
//Table中有两个字段(String name, Integer age)
Table table = ...
//把Table中的数据转成DataStream<Row>
DataStream<Row> dsRow = tableEnv.toAppendStream(table, Row.class);
//或者把Table中的数据转成DataStream<Tuple2>
TupleTypeInfo<Tuple2<String, Integer>> tupleType = new TupleTypeInfo<>(
  Types.STRING(),
  Types.INT());
DataStream<Tuple2<String, Integer>> dsTuple = tableEnv.toAppendStream(table, tupleType);
//将Table转化成Retract形式的DataStream<Row>
//一个Retract Stream的类型X为DataStream<Tuple2<Boolean, X>>
//Boolean字段指定了更改的类型
//true表示INSERT,false表示DELETE
DataStream<Tuple2<Boolean, Row>> retractStream = tableEnv.toRetractStream(table, Row.class);
```

Scala代码实现如下。

```scala
val env = StreamExecutionEnvironment.getExecutionEnvironment
val tableEnv = TableEnvironment.getTableEnvironment(env)
//Table中有两个字段(String name, Integer age)
val table: Table = ...
//把Table中的数据转成DataStream<Row>
val dsRow: DataStream[Row] = tableEnv.toAppendStream[Row](table)
//或者把Table中的数据转成DataStream<Tuple2>
val dsTuple: DataStream[(String, Int)] dsTuple = tableEnv.toAppendStream[(String, Int)](table)
val retractStream: DataStream[(Boolean, Row)] = tableEnv.toRetractStream[Row](table)
```

（2）把Table转化为DataSet。

Java代码实现如下。

```java
ExecutionEnvironment env = ExecutionEnvironment.getExecutionEnvironment();
BatchTableEnvironment tableEnv = TableEnvironment.getTableEnvironment(env);
//Table中有两个字段(String name, Integer age)
Table table = ...
//把Table中的数据转成DataSet<Row>
DataSet<Row> dsRow = tableEnv.toDataSet(table, Row.class);
//或者把Table中的数据转成DataSet<Tuple2>
TupleTypeInfo<Tuple2<String, Integer>> tupleType = new TupleTypeInfo<>(
  Types.STRING(),
  Types.INT());
DataSet<Tuple2<String, Integer>> dsTuple = tableEnv.toDataSet(table, tupleType);
```

Scala代码实现如下。

```scala
val env = StreamExecutionEnvironment.getExecutionEnvironment
val tableEnv = TableEnvironment.getTableEnvironment(env)
//Table中有两个字段(String name, Integer age)
val table: Table = ...
//把Table中的数据转成DataSet<Row>
val dsRow: DataSet[Row] = tableEnv.toDataSet[Row](table)
//或者把Table中的数据转成DataSet<Tuple2>
val dsTuple: DataSet[(String, Int)] = tableEnv.toDataSet[(String, Int)](table)
```

4.4.3 Table API和SQL的案例

1. 基于Flink Table API的案例

注意：在这里使用CsvTableSource，把结果直接打印到控制台。

需求：读取CSV文件中的内容，打印到控制台上。

CSV文件内容如下。

zs,15

ww,18

ls,20

Java代码实现如下。

```java
import org.apache.flink.api.common.typeinfo.TypeInformation;
import org.apache.flink.api.common.typeinfo.Types;
import org.apache.flink.streaming.api.datastream.DataStream;
import org.apache.flink.streaming.api.environment.StreamExecutionEnvironment;
import org.apache.flink.table.api.Table;
import org.apache.flink.table.api.TableEnvironment;
import org.apache.flink.table.api.java.StreamTableEnvironment;
import org.apache.flink.table.sources.CsvTableSource;
import org.apache.flink.table.sources.TableSource;

/**
 * Created by xuwei.tech.
 */
public class TableAPITest {
```

```java
    public static void main(String[] args)throws Exception {
        StreamExecutionEnvironment env = StreamExecutionEnvironment.getExecutionEnvironment();
        StreamTableEnvironment tableEnv = TableEnvironment.getTableEnvironment(env);
        //创建一个TableSource
        TableSource csvSource = new CsvTableSource("D:\\abc.csv",new String[]{"name","age"},new TypeInformation[]{Types.STRING, Types.INT});
        //注册一个TableSource，称为CvsTable
        tableEnv.registerTableSource("CsvTable", csvSource);

        Table csvTable = tableEnv.scan("CsvTable");
        Table csvResult = csvTable.select("name,age");
        DataStream<Student> csvStream = tableEnv.toAppendStream(csvResult, Student.class);
        csvStream.print().setParallelism(1);

        //执行任务
        env.execute("csvStream");

    }

    public static class Student {
        public String name;
        public int age;

        public Student() {}

        public Student(String name, int age) {
            this.name = name;
            this.age = age;
        }

        @Override
        public String toString() {
            return "name:" + name + ",age:" + age;
        }
    }
}
```

代码执行结果如下。

```
name:ls,age:20
name:zs,age:15
name:ww,age:18
```

Scala代码实现如下。

```scala
import org.apache.flink.api.common.typeinfo.TypeInformation
import org.apache.flink.api.scala.typeutils.Types
import org.apache.flink.streaming.api.scala.StreamExecutionEnvironment
import org.apache.flink.table.api.TableEnvironment
import org.apache.flink.table.sources.CsvTableSource

/**
  * Created by xuwei.tech.
  */
object TableTestScala {

  def main(args: Array[String]): Unit = {
    val sEnv = StreamExecutionEnvironment.getExecutionEnvironment
    val sTableEnv = TableEnvironment.getTableEnvironment(sEnv)
    //隐式转换
    import org.apache.flink.api.scala._

    //创建一个TableSource
    val csvSource = new CsvTableSource("D:\\abc.csv", Array[String]("name", "age"),
Array[TypeInformation[_]](Types.STRING, Types.INT))
    //注册一个TableSource,称为CvsTable
    sTableEnv.registerTableSource("CsvTable", csvSource)

    val csvTable = sTableEnv.scan("CsvTable")
    val csvResult = csvTable.select("name,age")

    val csvStream = sTableEnv.toAppendStream[Student](csvResult)
    csvStream.print.setParallelism(1)
    //执行任务
    sEnv.execute("csvStream")
  }

  case class Student(name: String,age: Int)

}
```

代码执行结果如下。

```
Student(zs,15)
Student(ls,20)
Student(ww,18)
```

2. 基于 Flink SQL 的案例

需求：读取 student.txt 文件中的单词并对其进行统计，计算每个单词出现的总次数，并把结果写入到 result.csv 文件中。

student.txt 文件内容如下。

zs,18

ls,20

ww,30

Java 代码实现如下。

```java
import org.apache.flink.api.common.functions.MapFunction;
import org.apache.flink.api.common.typeinfo.TypeInformation;
import org.apache.flink.api.common.typeinfo.Types;
import org.apache.flink.api.java.DataSet;
import org.apache.flink.api.java.ExecutionEnvironment;
import org.apache.flink.api.java.operators.DataSource;
import org.apache.flink.core.fs.FileSystem;
import org.apache.flink.table.api.Table;
import org.apache.flink.table.api.TableEnvironment;
import org.apache.flink.table.api.java.BatchTableEnvironment;
import org.apache.flink.table.sinks.CsvTableSink;

/**
 * Created by xuwei.tech.
 */
public class SQLTest {

    public static void main(String[] args) throws Exception{
        ExecutionEnvironment bEnv = ExecutionEnvironment.getExecutionEnvironment();
        BatchTableEnvironment bTableEnv = TableEnvironment.getTableEnvironment(bEnv);

        DataSource<String> dataSource = bEnv.readTextFile("D:\\student.txt");
        DataSet<Student> inputData = dataSource.map(new MapFunction<String, Student>() {
            @Override
            public Student map(String value) throws Exception {
                String[] splits = value.split(",");
                return new Student(splits[0], Integer.parseInt(splits[1]));
```

```java
            }
        });

        //将DataSet转换为Table
        Table table = bTableEnv.fromDataSet(inputData);
        //注册student表
        bTableEnv.registerTable("student",table);

        //执行Sql查询
        Table sqlQuery = bTableEnv.sqlQuery("select count(1),avg(age) from student");

        //创建CsvTableSink
        CsvTableSink csvTableSink = new CsvTableSink("D:\\result.csv", ",", 1, FileSystem.WriteMode.OVERWRITE);

        //注册TableSink
        bTableEnv.registerTableSink("csvOutputTable",new String[]{"count","avg_age"},new TypeInformation[]{Types.LONG,Types.INT},csvTableSink);

        //把结果数据添加到CsvTableSink中
        sqlQuery.insertInto("csvOutputTable");

        bEnv.execute("SQL-Batch");


    }
    //源数据
    public static class Student {
        public String name;
        public int age;

        public Student() {}

        public Student(String name, int age) {
            this.name = name;
            this.age = age;
        }

        @Override
        public String toString() {
            return "name:" + name + ",age:" + age;
        }
    }
}
```

代码执行结果被保存在result.csv文件中，内容如下。

```
3,22
```

Scala代码实现如下。

```scala
import org.apache.flink.api.common.typeinfo.TypeInformation
import org.apache.flink.api.scala.ExecutionEnvironment
import org.apache.flink.api.scala.typeutils.Types
import org.apache.flink.core.fs.FileSystem
import org.apache.flink.table.api.TableEnvironment
import org.apache.flink.table.sinks.CsvTableSink

/**
 * Created by xuwei.tech.
 */
object SQLTestScala {

  def main(args: Array[String]): Unit = {
    val bEnv = ExecutionEnvironment.getExecutionEnvironment
    val bTableEnv = TableEnvironment.getTableEnvironment(bEnv)

    //隐式转换
    import org.apache.flink.api.scala._

    val dataSource = bEnv.readTextFile("D:\\student.txt")
    val inputData = dataSource.map(line=>{
      val splits = line.split(",")
      val stu = new Student(splits(0),splits(1).toInt)
      stu
    })
    //将DataSet转换为Table
    val table = bTableEnv.fromDataSet(inputData)
    //注册student表
    bTableEnv.registerTable("student", table)
    //执行SQL查询
    val sqlQuery = bTableEnv.sqlQuery("select count(1),avg(age) from student")
    //创建CsvTableSink
    val csvTableSink = new CsvTableSink("D:\\result.csv", ",", 1, FileSystem.WriteMode.OVERWRITE)
    //注册TableSink
    bTableEnv.registerTableSink("csvOutputTable", Array[String]("count", "avg_age"), Array[TypeInformation[_]](Types.LONG, Types.INT), csvTableSink)
```

```
    //把结果数据添加到CsvTableSink中
    sqlQuery.insertInto("csvOutputTable")
    bEnv.execute("SQL-Batch")
  }

  case class Student(name: String,age: Int)

}
```

代码执行结果被保存在result.csv文件中,内容如下。

```
3,22
```

4.5　Flink支持的DataType分析

Flink支持Java和Scala中的大部分数据类型。

- Java Tuple和Scala Case Class。

- Java POJO：Java实体类。

- Primitive Type：默认支持Java和Scala基本数据类型。

- General Class Type：默认支持大多数Java和Scala Class。

- Hadoop Writable：支持Hadoop中实现了org.apache.Hadoop.Writable的数据类型。

- Special Type：比如Scala中的Either Option和Try。

4.6　Flink序列化分析

Flink自带了针对诸如Int、Long和String等标准类型的序列化器。

如果Flink无法实现序列化的数据类型,我们可以交给Avro和Kryo。

使用方法如下。

```
ExecutionEnvironment env = ExecutionEnvironment.getExecutionEnvironment();
```

- 使用 Avro 序列化：env.getConfig().enableForceAvro();。
- 使用 Kryo 序列化：env.getConfig().enableForceKryo();。
- 使用自定义序列化：env.getConfig().addDefaultKryoSerializer(Class<?> type, Class<? extends Serializer<?>> serializerClass);。

第5章
Flink 高级功能的使用

本章主要针对 Flink 中的高级特性进行分析，包括 Broadcast、Accumulator 和 Distributed Cache。

5.1　Flink Broadcast

在讲 Broadcast 之前需要区分一下 DataStream 中的 Broadcast（分区规则）和 Flink 中的 Broadcast（广播变量）功能。

1．DataStream Broadcast（分区规则）

分区规则是把元素广播给所有的分区，数据会被重复处理，类似于 Storm 中的 allGrouping。

```
DataStream.broadcast()
```

2．Flink Broadcast（广播变量）

广播变量允许编程人员在每台机器上保持一个只读的缓存变量，而不是传送变量的副本给 Task。广播变量创建后，它可以运行在集群中的任何 Function 上，而不需要多次传递给集群节点。另外请记住，不要修改广播变量，这样才能确保每个节点获取到的值都是一致的。

用一句话解释，Broadcast 可以理解为一个公共的共享变量。可以把一个 DataSet（数据集）广播出去，不同的 Task 在节点上都能够获取到它，这个数据集在每个节点上只会存在一份。如果不使用Broadcast，则在各节点的每个Task中都需要复制一份DataSet数据集，比较浪费内存（也就是一个节点中可能会存在多份DataSet数据）。

Broadcast的使用步骤如下。

（1）初始化数据。

```
DataSet<Integer> toBroadcast = env.fromElements(1, 2, 3)
```

（2）广播数据。

```
.withBroadcastSet(toBroadcast, "broadcastSetName");
```

（3）获取数据。

```
Collection<Integer>broadcastSet = getRuntimeContext().getBroadcastVariable("broadcastSetName");
```

在使用Broadcast的时候需要注意以下事项。

- 广播变量存在于每个节点的内存中，它的数据量不能太大，因为广播出去的数据常驻内存，除非程序执行结束。
- 广播变量在初始化广播以后不支持修改，这样才能保证每个节点的数据都是一致的。
- 如果多个算子需要使用同一份数据集，那么需要在对应的多个算子后面分别注册广播变量。
- 广播变量只能在Flink批处理程序中才可以使用。

需求：Flink从数据源中获取到用户的姓名，最终把用户的姓名和年龄信息打印出来。

分析：需要在中间的Map处理的时候获取用户的年龄信息，建议把用户的关系数据集使用广播变量进行处理。

注意：如果多个算子需要使用同一份数据集，那么需要在对应的多个算子后面分别注册广播变量。

Java代码实现如下。

```
package xuwei.tech.batch.batchAPI;

import org.apache.Flink.api.common.functions.MapFunction;
```

```java
import org.apache.Flink.api.common.functions.RichMapFunction;
import org.apache.Flink.api.java.DataSet;
import org.apache.Flink.api.java.ExecutionEnvironment;
import org.apache.Flink.api.java.operators.DataSource;
import org.apache.Flink.api.java.tuple.Tuple2;
import org.apache.Flink.configuration.Configuration;

import java.util.ArrayList;
import java.util.HashMap;
import java.util.List;

/**
 * Broadcast广播变量
 * Created by xuwei.tech
 */
public class BatchDemoBroadcast {

    public static void main(String[] args) throws Exception{

        //获取运行环境
        ExecutionEnvironment env = ExecutionEnvironment.getExecutionEnvironment();

        //1：准备需要广播的数据
        ArrayList<Tuple2<String, Integer>> broadData = new ArrayList<>();
        broadData.add(new Tuple2<>("zs",18));
        broadData.add(new Tuple2<>("ls",20));
        broadData.add(new Tuple2<>("ww",17));
        DataSet<Tuple2<String, Integer>> tupleData = env.fromCollection(broadData);

        //处理需要广播的数据,把数据集转换成Map类型,map中的key就是用户姓名,value就是用户年龄
        DataSet<HashMap<String, Integer>> toBroadcast = tupleData.map(new MapFunction<Tuple2<String, Integer>, HashMap<String, Integer>>() {
            @Override
            public HashMap<String, Integer> map(Tuple2<String, Integer> value) throws Exception {
                HashMap<String, Integer> res = new HashMap<>();
                res.put(value.f0, value.f1);
                return res;
            }
        });
```

```java
        //源数据
        DataSource<String> data = env.fromElements("zs", "ls", "ww");

        //注意：在这里使用RichMapFunction获取广播变量
        DataSet<String> result = data.map(new RichMapFunction<String, String>() {

            List<HashMap<String, Integer>> broadCastMap = new ArrayList<HashMap<String, Integer>>();
            HashMap<String, Integer> allMap = new HashMap<String, Integer>();

            /**
             * 这个方法只会执行一次
             * 可以在这里实现一些初始化的功能
             * 因此可以在open方法中获取广播变量数据
             */
            @Override
            public void open(Configuration parameters) throws Exception {
                super.open(parameters);
                //2：获取广播数据
                this.broadCastMap = getRuntimeContext().getBroadcastVariable("broadCastMapName");
                for (HashMap map : broadCastMap) {
                    allMap.putAll(map);
                }

            }

            @Override
            public String map(String value) throws Exception {
                Integer age = allMap.get(value);
                return value + "," + age;
            }
        }).withBroadcastSet(toBroadcast, "broadCastMapName");//3：执行广播数据的操作

        result.print();
    }
}
```

Scala代码实现如下。

```scala
package xuwei.tech.batch.batchAPI

import org.apache.Flink.api.common.functions.RichMapFunction
import org.apache.Flink.api.scala.ExecutionEnvironment
import org.apache.Flink.configuration.Configuration

import scala.collection.mutable.ListBuffer

/**
  * Broadcast广播变量
  * Created by xuwei.tech
  */
object BatchDemoBroadcastScala {

  def main(args: Array[String]): Unit = {

    val env = ExecutionEnvironment.getExecutionEnvironment

    import org.apache.Flink.api.scala._

    //1：准备需要广播的数据
    val broadData = ListBuffer[Tuple2[String,Int]]()
    broadData.append(("zs",18))
    broadData.append(("ls",20))
    broadData.append(("ww",17))

    //处理需要广播的数据
    val tupleData = env.fromCollection(broadData)
    val toBroadcastData = tupleData.map(tup=>{
      Map(tup._1->tup._2)
    })


    val text = env.fromElements("zs","ls","ww")


    val result = text.map(new RichMapFunction[String,String] {

      var listData: java.util.List[Map[String,Int]] = null
      var allMap  = Map[String,Int]()
```

```
      override def open(parameters: Configuration): Unit = {
        super.open(parameters)
      //2：获取广播数据
          this.listData = getRuntimeContext.getBroadcastVariable[Map[String,Int]]
("broadcastMapName")
        val it = listData.iterator()
        while (it.hasNext){
          val next = it.next()
          allMap = allMap.++(next)
        }
      }

      override def map(value: String) = {
        val age = allMap.get(value).get
        value+","+age
      }
    }).withBroadcastSet(toBroadcastData,"broadcastMapName")//3：执行广播数据的操作

    result.print()
    }
}
```

5.2　Flink Accumulator

Accumulator 即累加器，与 MapReduce 中 Counter 的应用场景差不多，都能很好地观察 Task 在运行期间的数据变化。可以在 Flink Job 的算子函数中使用累加器，但是只有在任务执行结束之后才能获得累加器的最终结果。

Counter 是一个具体的累加器实现，常用的 Counter 有 IntCounter、LongCounter 和 DoubleCounter。

累加器的使用步骤如下。

（1）创建累加器。

```
private IntCounter numLines = new IntCounter();
```

（2）注册累加器。

```
getRuntimeContext().addAccumulator("num-lines", this.numLines);
```

(3)使用累加器。

```
this.numLines.add(1);
```

(4)获取累加器的结果。

```
myJobExecutionResult.getAccumulatorResult("num-lines")
```

累加器需求:计算map函数中处理了多少数据。

注意:只有在任务执行结束后,才能获取累加器的值。

Java代码实现如下。

```
package xuwei.tech.batch.batchAPI;

import org.apache.Flink.api.common.JobExecutionResult;
import org.apache.Flink.api.common.accumulators.IntCounter;
import org.apache.Flink.api.common.functions.MapFunction;
import org.apache.Flink.api.common.functions.RichMapFunction;
import org.apache.Flink.api.java.DataSet;
import org.apache.Flink.api.java.ExecutionEnvironment;
import org.apache.Flink.api.java.operators.DataSource;
import org.apache.Flink.api.java.operators.MapOperator;
import org.apache.Flink.api.java.tuple.Tuple2;
import org.apache.Flink.configuration.Configuration;

import java.util.ArrayList;
import java.util.HashMap;
import java.util.List;

/**
 * 全局累加器
 * Created by xuwei.tech
 */
public class BatchDemoCounter {

    public static void main(String[] args) throws Exception{

        //获取运行环境
        ExecutionEnvironment env = ExecutionEnvironment.getExecutionEnvironment();
```

```java
            DataSource<String> data = env.fromElements("a", "b", "c", "d");

            DataSet<String> result = data.map(new RichMapFunction<String, String>() {

                //1：创建累加器
                private IntCounter numLines = new IntCounter();

                @Override
                public void open(Configuration parameters) throws Exception {
                    super.open(parameters);
                    //2：注册累加器
                    getRuntimeContext().addAccumulator("num-lines",this.numLines);

                }

                //int sum = 0;
                @Override
                public String map(String value) throws Exception {
                    //如果并行度为1，则使用普通的累加求和即可；如果设置多个并行度，则普通的累加求和
                    //结果就不准确了
                    //sum++;
                    //System.out.println("sum:"+sum);
                    this.numLines.add(1);
                    return value;
                }
            }).setParallelism(8);

            //result.print();

            result.writeAsText("d:\\data\\count10");

            JobExecutionResult jobResult = env.execute("counter");
            //3：获取累加器
            int num = jobResult.getAccumulatorResult("num-lines");
            System.out.println("num:"+num);

        }
    }
```

Scala代码实现如下。

```scala
package xuwei.tech.batch.batchAPI

import org.apache.Flink.api.common.accumulators.IntCounter
import org.apache.Flink.api.common.functions.RichMapFunction
import org.apache.Flink.api.scala.ExecutionEnvironment
import org.apache.Flink.configuration.Configuration

/**
  * Counter全局累加器
  * Created by xuwei.tech
  */
object BatchDemoCounterScala {

  def main(args: Array[String]): Unit = {

    val env = ExecutionEnvironment.getExecutionEnvironment

    import org.apache.Flink.api.scala._

    val data = env.fromElements("a","b","c","d")

    val res = data.map(new RichMapFunction[String,String] {
    //1：定义累加器
      val numLines = new IntCounter

      override def open(parameters: Configuration): Unit = {
        super.open(parameters)
        //2：注册累加器
        getRuntimeContext.addAccumulator("num-lines",this.numLines)
      }

      override def map(value: String) = {
        this.numLines.add(1)
        value
      }

    }).setParallelism(4)

    res.writeAsText("d:\\data\\count21")
    val jobResult = env.execute("BatchDemoCounterScala")
    //3：获取累加器
```

```
            val num = jobResult.getAccumulatorResult[Int]("num-lines")
            println("num:"+num)

    }

}
```

5.3 Flink Broadcast和Accumulator的区别

- Broadcast 允许程序员将一个只读的变量缓存在每台机器上，而不用在任务之间传递变量。广播变量可以进行共享，但是不可以进行修改。

- Accumulator 可以在不同任务中对同一个变量进行累加操作，但是只有在任务执行结束的时候才能获得累加器的最终结果。

5.4 Flink Distributed Cache

Flink 提供了一个分布式缓存（Distributed Cache），类似于Hadoop，可以使用户在并行函数中很方便地读取本地文件。

此缓存的工作机制为程序注册一个文件或者目录（本地或者远程文件系统，如HDFS或者S3），通过 ExecutionEnvironment 注册缓存文件并为它起一个名称。当程序执行时，Flink 自动将文件或者目录复制到所有TaskManager 节点的本地文件系统，用户可以通过这个指定的名称查找文件或者目录，然后从TaskManager 节点的本地文件系统访问它。

Flink 分布式缓存的使用步骤。

（1）注册一个文件。

```
env.registerCachedFile("hdfs:///path/to/your/file", "hdfsFile")
```

（2）访问数据。

```
File myFile = getRuntimeContext().getDistributedCache().getFile("hdfsFile");
```

分布式缓存案例的Java代码实现如下。

```java
package xuwei.tech.batch.batchAPI;

import org.apache.commons.io.FileUtils;
import org.apache.Flink.api.common.functions.MapFunction;
import org.apache.Flink.api.common.functions.RichMapFunction;
import org.apache.Flink.api.java.DataSet;
import org.apache.Flink.api.java.ExecutionEnvironment;
import org.apache.Flink.api.java.operators.DataSource;
import org.apache.Flink.api.java.operators.MapOperator;
import org.apache.Flink.api.java.tuple.Tuple2;
import org.apache.Flink.configuration.Configuration;

import java.io.File;
import java.util.ArrayList;
import java.util.HashMap;
import java.util.List;

/**
 * Distributed Cache
 * Created by xuwei.tech
 */
public class BatchDemoDisCache {

    public static void main(String[] args) throws Exception{

        //获取运行环境
        ExecutionEnvironment env = ExecutionEnvironment.getExecutionEnvironment();

        //1：注册一个文件，可以使用HDFS或者S3上的文件
        env.registerCachedFile("d:\\data\\file\\a.txt","a.txt");

        DataSource<String> data = env.fromElements("a", "b", "c", "d");

        DataSet<String> result = data.map(new RichMapFunction<String, String>() {
            private ArrayList<String> dataList = new ArrayList<String>();

            @Override
            public void open(Configuration parameters) throws Exception {
                super.open(parameters);
                //2：使用文件
                File myFile = getRuntimeContext().getDistributedCache().getFile("a.txt");
                List<String> lines = FileUtils.readLines(myFile);
```

```
                for (String line : lines) {
                    this.dataList.add(line);
                    System.out.println("line:" + line);
                }
            }

            @Override
            public String map(String value) throws Exception {
                //在这里就可以使用dataList
                return value;
            }
        });

        result.print();
    }
}
```

Scala代码实现如下。

```
package xuwei.tech.batch.batchAPI

import org.apache.commons.io.FileUtils
import org.apache.Flink.api.common.functions.RichMapFunction
import org.apache.Flink.api.scala.ExecutionEnvironment
import org.apache.Flink.configuration.Configuration

/**
  * Distributed Cache
  * Created by xuwei.tech
  */
object BatchDemoDisCacheScala {

  def main(args: Array[String]): Unit = {

    val env = ExecutionEnvironment.getExecutionEnvironment

    import org.apache.Flink.api.scala._


    //1：注册文件
    env.registerCachedFile("d:\\data\\file\\a.txt","b.txt")
```

```scala
    val data = env.fromElements("a","b","c","d")

    val result = data.map(new RichMapFunction[String,String] {

      override def open(parameters: Configuration): Unit = {
        super.open(parameters)
        //2：使用文件
        val myFile = getRuntimeContext.getDistributedCache.getFile("b.txt")
        val lines = FileUtils.readLines(myFile)
        val it = lines.iterator()
        while (it.hasNext){
          val line = it.next();
          println("line:"+line)
        }
      }
      override def map(value: String) = {
        value
      }
    })

    result.print()

  }
}
```

第6章
Flink State 管理与恢复

本章主要针对 Flink State(状态)进行分析,包含状态的管理和恢复,以及 Flink 中的任务重启策略。

6.1 State

前面 word count 的例子没有包含状态管理。如果一个 Task 在处理过程中挂掉了,那么它在内存中的状态都会丢失,所有的数据都需要重新计算。从容错和消息处理的语义(At-least-once 和 Exactly-once)上来说,Flink 引入了 State 和 CheckPoint。

这两个概念的区别如下。

- State 一般指一个具体的 Task/Operator 的状态,State 数据默认保存在 Java 的堆内存中。

- 而 CheckPoint(可以理解为 CheckPoint 是把 State 数据持久化存储了)则表示了一个 Flink Job 在一个特定时刻的一份全局状态快照,即包含了所有 Task/Operator 的状态。

注意:Task 是 Flink 中执行的基本单位,Operator 是算子(Transformation)。

State 可以被记录,在失败的情况下数据还可以恢复。Flink 中有以下两种基本类型的 State。

- Keyed State。

- Operator State。

Keyed State 和 Operator State 以两种形式存在。

- 原始状态（Raw State）：由用户自行管理状态具体的数据结构，框架在做 CheckPoint 的时候，使用 byte[] 读写状态内容，对其内部数据结构一无所知。
- 托管状态（Managed State）：由 Flink 框架管理的状态。

通常在 DataStream 上推荐使用托管状态，当实现一个用户自定义的 Operator 时使用到原始状态。

6.1.1 Keyed State

Keyed State，顾名思义就是基于 KeyedStream 上的状态，这个状态是跟特定的 Key 绑定的。KeyedStream 流上的每一个 Key，都对应一个 State。

stream.keyBy(……)这个代码会返回一个 KeyedStream 对象。

Flink 针对 Keyed State 提供了以下可以保存 State 的数据结构。

- ValueState<T>：类型为 T 的单值状态，这个状态与对应的 Key 绑定，是最简单的状态。它可以通过 update 方法更新状态值，通过 value() 方法获取状态值。
- ListState<T>：Key 上的状态值为一个列表，这个列表可以通过 add 方法往列表中附加值，也可以通过 get() 方法返回一个 Iterable<T> 来遍历状态值。
- ReducingState<T>：每次调用 add 方法添加值的时候，会调用用户传入的 reduceFunction，最后合并到一个单一的状态值。
- MapState<UK, UV>：状态值为一个 Map，用户通过 put 或 putAll 方法添加元素。

需要注意的是，以上所述的 State 对象，仅仅用于与状态进行交互（更新、删除、清空等），而真正的状态值有可能存在于内存、磁盘或者其他分布式存储系统中，相当于我们只是持有了这个状态的句柄。

案例代码如下。

```
public class CountWindowAverage extends RichFlatMapFunction<Tuple2
<Long, Long>, Tuple2<Long, Long>> {

    /**
     * ValueState 状态句柄，第一个字段为 count，第二个字段为 sum
```

```java
         */
        private transient ValueState<Tuple2<Long, Long>> sum;

        @Override
        public void flatMap(Tuple2<Long, Long> input, Collector<Tuple2<Long, Long>> out)
throws Exception {

            // 获取当前状态值
            Tuple2<Long, Long> currentSum = sum.value();

            // 更新
            currentSum.f0 += 1;
            currentSum.f1 += input.f1;

            // 更新状态值
            sum.update(currentSum);

            // 如果count >=2清空状态值,重新计算
            if (currentSum.f0 >= 2) {
                out.collect(new Tuple2<>(input.f0, currentSum.f1 / currentSum.f0));
                sum.clear();
            }
        }

        @Override
        public void open(Configuration config) {
            ValueStateDescriptor<Tuple2<Long, Long>> descriptor =
                    new ValueStateDescriptor<>(
                            "average", // 状态名称
                            TypeInformation.of(new TypeHint<Tuple2<Long, Long>>() {}), // 状态类型
                            Tuple2.of(0L, 0L)); // 状态默认值
            sum = getRuntimeContext().getState(descriptor);
        }
    }

// 算子计算处理
env.fromElements(Tuple2.of(1L, 3L), Tuple2.of(1L, 5L), Tuple2.of(1L, 7L), Tuple2.
of(1L, 4L), Tuple2.of(1L, 2L))
        .keyBy(0)
        .flatMap(new CountWindowAverage())
        .print();

// 最终打印的结果是 (1,4) 和 (1,5)
```

6.1.2 Operator State

Operator State 与 Key 无关，而是与 Operator 绑定，整个 Operator 只对应一个 State。

Flink 针对 Operator State 提供了以下可以保存 State 的数据结构。

```
ListState<T>
```

举例来说，Flink 中的 Kafka Connector 就使用了 Operator State，它会在每个 Connector 实例中，保存该实例消费 Topic 的所有 (partition, offset) 映射，如图 6.1 所示。

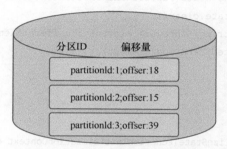

图 6.1　Kafka 中 Topic 的消费信息

案例代码如下。

```java
public class BufferingSink
        implements SinkFunction<Tuple2<String, Integer>>,
                CheckpointedFunction {

    private final int threshold;

    private transient ListState<Tuple2<String, Integer>> checkpointedState;

    private List<Tuple2<String, Integer>> bufferedElements;

    public BufferingSink(int threshold) {
        this.threshold = threshold;
        this.bufferedElements = new ArrayList<>();
    }

    @Override
    public void invoke(Tuple2<String, Integer> value) throws Exception {
        bufferedElements.add(value);
```

```java
        if (bufferedElements.size() == threshold) {
            for (Tuple2<String, Integer> element: bufferedElements) {
                // send it to the sink
            }
            bufferedElements.clear();
        }
    }

    @Override
    public void snapshotState(FunctionSnapshotContext context) throws Exception {
        checkpointedState.clear();
        for (Tuple2<String, Integer> element : bufferedElements) {
            checkpointedState.add(element);
        }
    }

    @Override
    public void initializeState(FunctionInitializationContext context) throws Exception {
        ListStateDescriptor<Tuple2<String, Integer>> descriptor =
            new ListStateDescriptor<>(
                "buffered-elements",
                TypeInformation.of(new TypeHint<Tuple2<String, Integer>>() {}));

        checkpointedState = context.getOperatorStateStore().getListState(descriptor);

        if (context.isRestored()) {
            for (Tuple2<String, Integer> element : checkpointedState.get()) {
                bufferedElements.add(element);
            }
        }
    }
}
```

6.2 State 的容错

当程序出现问题需要恢复 Sate 数据的时候，只有程序提供支持才可以实现 State 的容错。

State 的容错需要依靠 CheckPoint 机制，这样才可以保证 Exactly-once 这种语义，但是注意，它只能保证 Flink 系统内的 Exactly-once，比如 Flink 内置支持的算子。

针对 Source 和 Sink 组件，如果想要保证 Exactly-once 的话，则这些组件本身应支持这种语义。

下面我们分析 State 容错中的 CheckPoint 机制。

1. 生成快照

Flink 通过 CheckPoint 机制可以实现对 Source 中的数据和 Task 中的 State 数据进行存储。如图 6.2 所示。

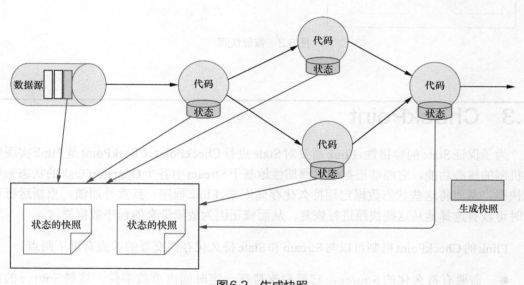

图 6.2 生成快照

2. 恢复快照

Flink 还可以通过 Restore 机制来恢复之前 CheckPoint 快照中保存的 Source 数据和 Task 中的 State 数据。如图 6.3 所示。

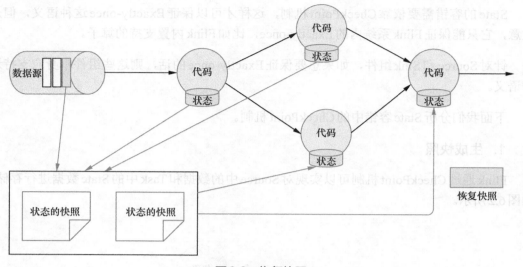

图6.3 恢复快照

6.3 CheckPoint

为了保证State的容错性，Flink需要对State进行CheckPoint。CheckPoint是Flink实现容错机制的核心功能，它能够根据配置周期性地基于Stream中各个Operator/Task的状态来生成快照，从而将这些状态数据定期持久化存储下来。Flink程序一旦意外崩溃，重新运行程序时可以有选择地从这些快照进行恢复，从而修正因为故障带来的程序数据异常。

Flink的CheckPoint机制可以与Stream和State持久化存储交互的前提有以下两点。

- 需要有持久化的Source，它需要支持在一定时间内重放事件，这种Source的典型例子就是持久化的消息队列（如Apache Kafka、RabbitMQ等）或文件系统（如HDFS、S3、GFS等）。
- 需要有用于State的持久化存储介质，比如分布式文件系统（如HDFS、S3、GFS等）。

默认情况下，CheckPoint功能是Disabled(禁用)的，使用时需要先开启它。

通过如下代码即可开启。

```
env.enableCheckpointing(1000);
```

完整的参考代码如下。

```
StreamExecutionEnvironment env = StreamExecutionEnvironment.getExecutionEnvironment();
// 每隔1000 ms启动一个检查点(设置CheckPoint的周期)
env.enableCheckpointing(1000);
// 高级选项:
// 设置模式为Exactly-once (这是默认值)
env.getCheckpointConfig().setCheckpointingMode(CheckpointingMode.EXACTLY_ONCE);
// 确保检查点之间有至少500 ms的间隔(CheckPoint最小间隔)
env.getCheckpointConfig().setMinPauseBetweenCheckpoints(500);
// 检查点必须在1min内完成,或者被丢弃(CheckPoint的超时时间)
env.getCheckpointConfig().setCheckpointTimeout(60000);
// 同一时间只允许操作一个检查点
env.getCheckpointConfig().setMaxConcurrentCheckpoints(1);
// 表示一旦Flink处理程序被cancel后,会保留CheckPoint数据,以便根据实际需要恢复到指定的
CheckPoint
env.getCheckpointConfig().enableExternalizedCheckpoints(ExternalizedCheckpointClean
up.RETAIN_ON_CANCELLATION);
```

注意:enableExternalizedCheckpoints()方法中可以接收以下两个参数。

- ExternalizedCheckpointCleanup.RETAIN_ON_CANCELLATION:表示一旦Flink处理程序被cancel后,会保留CheckPoint数据,以便根据实际需要恢复到指定的CheckPoint。

- ExternalizedCheckpointCleanup.DELETE_ON_CANCELLATION:表示一旦Flink处理程序被cancel后,会删除CheckPoint数据,只有Job执行失败的时候才会保存CheckPoint。

当CheckPoint机制开启之后,默认的CheckPointMode是Exactly-once,CheckPointMode有两种选项:Exactly-once和At-least-once。

Exactly-once对于大多数应用来说是合适的,At-least-once可能用在某些延迟超低的应用程序(始终延迟为几毫秒)上。

6.4 StateBackend

默认情况下,State会保存在TaskManager的内存中,CheckPoint会存储在JobManager的内存中。State和CheckPoint的存储位置取决于StateBackend的配置。Flink一共提供了3种StateBackend。

1. MemoryStateBackend

State 数据保存在 Java 堆内存中，执行 CheckPoint 的时候，会把 State 的快照数据保存到 JobManager 的内存中。基于内存的 StateBackend 在生产环境下不建议使用。

2. FsStateBackend

State 数据保存在 TaskManager 的内存中，执行 CheckPoint 的时候，会把 State 的快照数据保存到配置的文件系统中，可以使用 HDFS 等分布式文件系统。

3. RocksDBStateBackend

RocksDB 跟上面的都略有不同，它会在本地文件系统中维护状态，State 会直接写入本地 RocksDB 中。同时它需要配置一个远端的 FileSystem URI（一般是 HDFS），在进行 CheckPoint 的时候，会把本地的数据直接复制到远端的 FileSystem 中。Fail Over（故障切换）的时候直接从远端的 Filesystem 中恢复数据到本地。RocksDB 克服了 State 受内存限制的缺点，同时又能够持久化到远端文件系统中，推荐在生产中使用。

如何修改 Flink 的 State 和 CheckPoint 存储的位置呢？有以下两种方式。

1. 单任务调整

修改当前任务的代码。

```
env.setStateBackend(new FsStateBackend("hdfs://hadoop100:9000/flink/checkpoints"));
```

可以给 setStateBackend() 方法传递 new MemoryStateBackend()，也可以给 setStateBackend() 方法传递 new RocksDBStateBackend(filebackend, true)。

注意：在使用 RocksDBStateBackend 这个类的时候需要引入第三方依赖。

```xml
<dependency>
    <groupId>org.apache.flink</groupId>
    <artifactId>flink-statebackend-rocksdb_2.11</artifactId>
    <version>1.6.1</version>
</dependency>
```

2. 全局调整

需要修改 flink-conf.yaml 配置文件，主要修改下面两个参数。

- state.backend: filesystem。

- state.checkpoints.dir: hdfs://namenode:9000/flink/checkpoints。

注意：state.backend的值可以是下面这3种。

- jobmanager表示使用MemoryStateBackend。
- filesystem表示使用FsStateBackend。
- rocksdb表示使用RocksDBStateBackend。

默认情况下，如果设置了CheckPoint选项，则Flink只保留最近成功生成的1个CheckPoint，而当Flink程序失败时，可以通过最近的CheckPoint来进行恢复。但是，如果希望保留多个CheckPoint，并能够根据实际需要选择其中一个进行恢复，就会更加灵活。

比如，我们发现最近4h的数据记录处理有问题，希望将整个状态还原到4h之前。Flink可以支持保留多个CheckPoint，需要在Flink的配置文件conf/flink-conf.yaml中添加如下配置，指定最多可以保存的CheckPoint的个数。

```
state.checkpoints.num-retained: 20
```

这样设置以后就可以查看到对应的CheckPoint在HDFS上的文件目录。

可以执行下面命令查看。

```
hdfs dfs -ls hdfs://hadoop100:9000/flink/checkpoints
```

如果希望回退到某个CheckPoint点，只需要指定对应的某个CheckPoint路径即可实现。

比如，某Flink程序异常失败，或者最近一段时间内数据处理错误，就可以将程序通过某一个CheckPoint点进行恢复。

```
bin/flink run -s hdfs://hadoop100:9000/flink/checkpoints/467e17d2cc343e6c56255d222bae3421/chk-56/_metadata flink-job.jar
```

程序正常运行后，还会按照CheckPoint配置运行，继续生成CheckPoint数据。

6.5　Restart Strategy

Flink支持不同的Restart Strategy（重启策略），以便在故障发生时控制作业重启。集群在启动时会伴随一个默认的重启策略，在没有定义具体重启策略时会使用该默认策略；如果在任务提交时指定了一个重启策略，该策略会覆盖集群的默认策略。

默认的重启策略是通过Flink的配置文件flink-conf.yaml中的restart-strategy参数指定的。

常用的重启策略有以下3种。

- 固定间隔（Fixed delay）。
- 失败率（Failure rate）。
- 无重启（No restart）。

如果没有启用CheckPoint，则使用无重启策略。如果启用了CheckPoint，但没有配置重启策略，则使用固定间隔策略，其中Integer.MAX_VALUE参数是允许尝试重启的次数。

重启策略可以在flink-conf.yaml中配置，这属于全局配置，也可以在某个任务代码中动态指定，且只对这个任务有效，会覆盖全局的配置。

下面来详细分析一下这3种重启策略。

1. 固定间隔

全局配置，修改flink-conf.yaml中的参数。

```
restart-strategy: fixed-delay
restart-strategy.fixed-delay.attempts: 3
restart-strategy.fixed-delay.delay: 10 s
```

在任务代码中做如下配置。

```
env.setRestartStrategy(RestartStrategies.fixedDelayRestart(
  3, // 尝试重启的次数
  Time.of(10, TimeUnit.SECONDS) // 间隔
));
```

2. 失败率

全局配置，修改flink-conf.yaml中的参数。

```
restart-strategy: failure-rate
restart-strategy.failure-rate.max-failures-per-interval: 3
restart-strategy.failure-rate.failure-rate-interval: 5 min
restart-strategy.failure-rate.delay: 10 s
```

在任务代码中做如下配置。

```
env.setRestartStrategy(RestartStrategies.failureRateRestart(
    3, // 一个时间段内的最大失败次数
    Time.of(5, TimeUnit.MINUTES), // 衡量失败次数的是时间段
    Time.of(10, TimeUnit.SECONDS) // 间隔
));
```

3. 无重启

全局配置，修改 flink-conf.yaml 中的参数。

```
restart-strategy: none
```

在任务代码中做如下配置。

```
env.setRestartStrategy(RestartStrategies.noRestart());
```

6.6 SavePoint

Flink 通过 SavePoint 功能可以升级程序，然后继续从升级前的那个点开始执行计算，保证数据不中断。SavePoint 可以生成全局、一致性的快照，也可以保存数据源、Offset、Operator 操作状态等信息，还可以从应用在过去任意做了 SavePoint 的时刻开始继续执行。

那么这个 SavePoint 和我们前面说的 CheckPoint 有什么区别呢？

1. CheckPoint

应用定时触发，用于保存状态，它会过期，在内部应用失败重启的时候使用。

2. SavePoint

用户手动执行，是指向 CheckPoint 的指针，它不会过期，一般在升级的情况下使用。

注意：为了能够在作业的不同版本之间以及 Flink 的不同版本之间顺利升级，强烈推荐程序员通过 UID（String）方法手动给算子赋予 ID，这些 ID 将用于确定每一个算子的状态范围。如果不手动给各算子指定 ID，则会由 Flink 自动给每个算子生成一个 ID。只要这些 ID 没有改变，就能从保存点（SavePoint）恢复程序。而这些自动生成的 ID 依赖于程序的结构，并且对代码的更改是很敏感的。因此，强烈建议用户手动设置 ID。如图 6.4 所示的代码。

```
DataStream<String> stream = env.
  // Stateful source (e.g. Kafka) with ID
  .addSource(new StatefulSource())
  .uid("source-id") // ID for the source operator
  .shuffle()
  // Stateful mapper with ID
  .map(new StatefulMapper())
  .uid("mapper-id") // ID for the mapper
  // Stateless printing sink
  .print(); // Auto-generated ID
```

图6.4 给算子赋予ID

如何配置开启SavePoint功能呢？

（1）在flink-conf.yaml中配置SavePoint存储的位置。

这不是必须设置的，但是在设置后，如果要创建指定Job的SavePoint，可以不用在手动执行命令时指定SavePoint的位置。

```
state.savepoints.dir: hdfs://hadoop100:9000/Flink/savepoints
```

（2）触发一个SavePoint（直接触发或者在cancel的时候触发）。

直接触发SavePoint。

```
bin/flink savepoint jobId [targetDirectory] [-yid yarnAppId]
```

在调用cancel的时候触发SavePoint。

```
bin/flink cancel -s [targetDirectory] jobId [-yid yarnAppId]
```

注意：以上两种方式，针对Flink on YARN模式需要指定-yid参数。

（3）从指定的SavePoint启动Job。

```
bin/flink run -s savepointPath [runArgs]
```

第 7 章
Flink 窗口详解

本章主要针对 Flink 窗口（Window）进行分析，包括 Flink 中提供的常见 Window，以及 Window 的聚合操作。

7.1 Window

Flink 认为 Batch 是 Streaming 的一个特例，因此 Flink 底层引擎是一个流式引擎，在上面实现了流处理和批处理。而 Window 就是从 Streaming 到 Batch 的桥梁。

通常来讲，Window 就是用来对一个无限的流设置一个有限的集合，从而在有界的数据集上进行操作的一种机制。

比如，对流中的所有元素进行计数是不可能的，因为通常流是无限的（无界的）。因此，流上的聚合需要由 Window 来划定范围，比如"计算过去的 5min"或者"最后 100 个元素的和"。

Window 可以由时间（Time Window）（如每 30s）或者数据（Count Window）（如每 100 个元素）驱动。DataStream API 提供了 Time 和 Count 的 Window。同时，由于某些特殊的需要，DataStream API 也提供了定制化的 Window 操作，供用户自定义 Window。如图 7.1 所示。

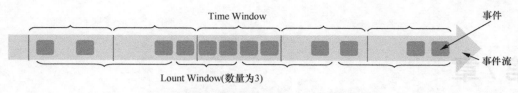

图7.1 Time Window和Count Window

7.2 Window的使用

下面来详细分析Window中的Time Window、Count Window以及自定义Window。

Window根据类型可以分为两种。

- **Tumbling Window**：滚动窗口，表示窗口内的数据没有重叠，如图7.2所示。

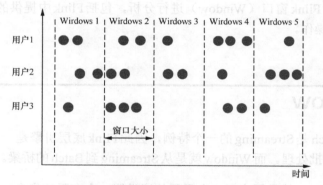

图7.2 Tumbling Window

- **Sliding Window**：滑动窗口，表示窗口内的数据有重叠，如图7.3所示。

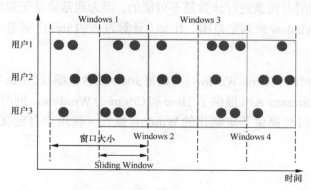

图7.3 Sliding Window

Window 类型之间的关系如图 7.4 所示。

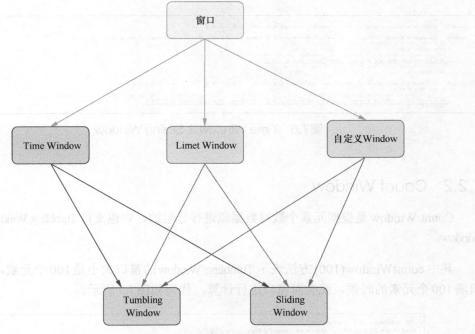

图 7.4　Window 类型之间的关系

7.2.1　Time Window

Time Window 是根据时间对数据流进行分组的，它支持 Tumbling Window 和 Sliding Window。

其中 timeWindow(Time.minutes(1)) 方法表示 Tumbling 窗口的窗口大小为 1min，对每 1min 内的数据进行聚合计算，代码如图 7.5 所示。

```
// Stream of (sensorId, carCnt)
val vehicleCnts: DataStream[(Int, Int)] = ...

val tumblingCnts: DataStream[(Int, Int)] = vehicleCnts
  // key stream by sensorId
  .keyBy(0)
  // tumbling time window of 1 minute length
  .timeWindow(Time.minutes(1))
  // compute sum over carCnt
  .sum(1)

val slidingCnts: DataStream[(Int, Int)] = vehicleCnts
  .keyBy(0)
  // sliding time window of 1 minute length and 30 secs trigger interval
  .timeWindow(Time.minutes(1), Time.seconds(30))
  .sum(1)
```

图 7.5　Time Window 之 Tumbling Window

timeWindow(Time.minutes(1),Time.seconds(30))方法表示Sliding Window的窗口大小为1min，滑动间隔为30s。就是每隔30s计算最近1min内的数据，代码如图7.6所示。

```
// Stream of (sensorId, carCnt)
val vehicleCnts: DataStream[(Int, Int)] = ...
val tumblingCnts: DataStream[(Int, Int)] = vehicleCnts
  // key stream by sensorId
  .keyBy(0)
  // tumbling time window of 1 minute length
  .timeWindow(Time.minutes(1))
  // compute sum over carCnt
  .sum(1)
val slidingCnts: DataStream[(Int, Int)] = vehicleCnts
  .keyBy(0)
  // sliding time window of 1 minute length and 30 secs trigger interval
  .timeWindow(Time.minutes(1), Time.seconds(30))
  .sum(1)
```

图7.6　Time Window之Sliding Window

7.2.2　Count Window

Count Window是根据元素个数对数据流进行分组的，它也支持Tumbling Window和Sliding Window。

其中countWindow(100)方法表示Tumbling Window的窗口大小是100个元素，当窗口中填满100个元素的时候，就会对窗口进行计算，代码如图7.7所示。

```
// Stream of (sensorId, carCnt)
val vehicleCnts: DataStream[(Int, Int)] = ...
val tumblingCnts: DataStream[(Int, Int)] = vehicleCnts
  // key stream by sensorId
  .keyBy(0)
  // tumbling count window of 100 elements size
  .countWindow(100)
  // compute the carCnt sum
  .sum(1)
val slidingCnts: DataStream[(Int, Int)] = vehicleCnts
  .keyBy(0)
  // sliding count window of 100 elements size and 10 elements trigger interval
  .countWindow(100, 10)
  .sum(1)
```

图7.7　Count Window之Tumbling Window

countWindow(100,10)方法表示Sliding Window的窗口大小是100个元素，滑动的间隔为10个元素，也就是说每新增10个元素就会对前面100个元素计算一次，代码如图7.8所示。

```
// Stream of (sensorId, carCnt)
val vehicleCnts: DataStream[(Int, Int)] = ...
val tumblingCnts: DataStream[(Int, Int)] = vehicleCnts
  // key stream by sensorId
  .keyBy(0)
  // tumbling count window of 100 elements size
  .countWindow(100)
  // compute the carCnt sum
  .sum(1)
val slidingCnts: DataStream[(Int, Int)] = vehicleCnts
  .keyBy(0)
  // sliding count window of 100 elements size and 10 elements trigger interval
  .countWindow(100, 10)
  .sum(1)
```

图7.8　Count Window之Sliding Window

7.2.3 自定义 Window

自定义 Window 可以分为两种：一种是基于 Key 的 Window，一种是不基于 Key 的 Window，如图 7.9 所示。

- .keyBy(...).widow(...)：属于基于 Key 的 Window，会先对窗口中的数据进行分组，然后再分组。

- .windowAll(...)：属于不基于 Key 的 Window，会对窗口所有数据进行聚合。

```
基于Key的 Window

stream
       .keyBy(...)                  <- keyed versus non-keyed windows
       .window(...)                 <- required: "assigner"
      [.trigger(...)]               <- optional: "trigger" (else default trigger)
      [.evictor(...)]               <- optional: "evictor" (else no evictor)
      [.allowedLateness(...)]       <- optional: "lateness" (else zero)
      [.sideOutputLateData(...)]    <- optional: "output tag" (else no side output for late data)
       .reduce/aggregate/fold/apply()  <- required: "function"
      [.getSideOutput(...)]         <- optional: "output tag"

不基于Key的 Window

stream
       .windowAll(...)              <- required: "assigner"
      [.trigger(...)]               <- optional: "trigger" (else default trigger)
      [.evictor(...)]               <- optional: "evictor" (else no evictor)
      [.allowedLateness(...)]       <- optional: "lateness" (else zero)
      [.sideOutputLateData(...)]    <- optional: "output tag" (else no side output for late data)
       .reduce/aggregate/fold/apply()  <- required: "function"
      [.getSideOutput(...)]         <- optional: "output tag"
```

图 7.9 自定义 Window

TimeWindow 和 Count Window 是针对 Window 的封装，代码如图 7.10 所示。

```
public WindowedStream<T, KEY, TimeWindow> timeWindow(Time size) {
    if (environment.getStreamTimeCharacteristic() == TimeCharacteristic.ProcessingTime) {
        return window(TumblingProcessingTimeWindows.of(size));
    } else {
        return window(TumblingEventTimeWindows.of(size));
    }
}
```

图 7.10 Time Window 底层实现

相对应的，也可以通过 timeWindowAll 来实现，这个方法对 windowAll 进行了封装，如图 7.11 所示。

```
public AllWindowedStream<T, TimeWindow> timeWindowAll(Time size) {
    if (environment.getStreamTimeCharacteristic() == TimeCharacteristic.ProcessingTime) {
        return windowAll(TumblingProcessingTimeWindows.of(size));
    } else {
        return windowAll(TumblingEventTimeWindows.of(size));
    }
}
```

图7.11 timeWindowAll底层实现

7.3 Window聚合分类

Window聚合操作分为两种：一种是增量聚合，一种是全量聚合。增量聚合是指窗口每进入一条数据就计算一次，而全量聚合是指在窗口触发的时候才会对窗口内的所有数据进行一次计算。

7.3.1 增量聚合

常见的增量聚合函数如下。

- reduce(reduceFunction)。
- aggregate(aggregateFunction)。
- sum()。
- min()。
- max()。

下面来看一个增量聚合的例子：累加求和，如图7.12所示。

- 第一次进来一条数据8，则立刻进行累加求和，结果为8。
- 第二次进来一条数据12，则立刻进行累加求和，结果为20。
- 第三次进来一条数据7，则立刻进行累加求和，结果为27。
- 第四次进来一条数据10，则立刻进行累加求和，结果为37。

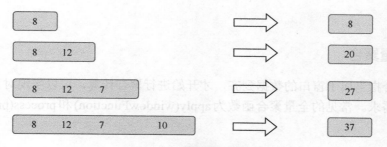

图7.12 增量聚合案例

下面分析reduce函数的使用步骤,代码如图7.13所示。

```
DataStream<Tuple2<String, Long>> input = ...;

input
    .keyBy(<key selector>)
    .window(<window assigner>)
    .reduce(new ReduceFunction<Tuple2<String, Long>> {
        public Tuple2<String, Long> reduce(Tuple2<String, Long> v1, Tuple2<String, Long> v2) {
            return new Tuple2<>(v1.f0, v1.f1 + v2.f1);
        }
    });
```

图7.13 reduce增量聚合函数案例

接下来分析aggregate函数的使用步骤,代码如图7.14所示。

```
private static class AverageAggregate
    implements AggregateFunction<Tuple2<String, Long>, Tuple2<Long, Long>, Double> {
    @Override
    public Tuple2<Long, Long> createAccumulator() {
        return new Tuple2<>(0L, 0L);
    }

    @Override
    public Tuple2<Long, Long> add(Tuple2<String, Long> value, Tuple2<Long, Long> accumulator) {
        return new Tuple2<>(accumulator.f0 + value.f1, accumulator.f1 + 1L);
    }

    @Override
    public Double getResult(Tuple2<Long, Long> accumulator) {
        return ((double) accumulator.f0) / accumulator.f1;
    }

    @Override
    public Tuple2<Long, Long> merge(Tuple2<Long, Long> a, Tuple2<Long, Long> b) {
        return new Tuple2<>(a.f0 + b.f0, a.f1 + b.f1);
    }
}

DataStream<Tuple2<String, Long>> input = ...;

input
    .keyBy(<key selector>)
    .window(<window assigner>)
    .aggregate(new AverageAggregate());
```

图7.14 aggregate增量聚合函数案例

7.3.2 全量聚合

全量聚合指当属于窗口的数据到齐,才开始进行聚合计算,可以实现对窗口内的数据进行排序等需求。常见的全量聚合函数为apply(windowFunction)和process(processWindowFunction)。

注意:processWindowFunction比windowFunction提供了更多的Context(上下文)信息。

下面来看一个全量聚合的例子:求最大值,如图7.15所示。

- 第一次进来一条数据8。
- 第二次进来一条数据12。
- 第三次进来一条数据7。
- 第四次进来一条数据10,此时窗口触发,才会对窗口内的数据进行排序,获取最大值。

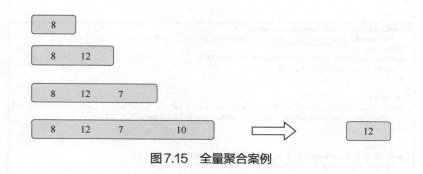

图7.15 全量聚合案例

下面来分析apply函数的使用步骤,代码如图7.16所示。

接下来看process函数的使用步骤,代码如图7.17所示。

在ProcessWindowFunction中提供了一个Context(上下文)对象,里面提供了针对State的一些操作。

7.3 Window聚合分类

```java
public interface WindowFunction<IN, OUT, KEY, W extends Window> extends Function, Serializable {

    /**
     * Evaluates the window and outputs none or several elements.
     *
     * @param key The key for which this window is evaluated.
     * @param window The window that is being evaluated.
     * @param input The elements in the window being evaluated.
     * @param out A collector for emitting elements.
     *
     * @throws Exception The function may throw exceptions to fail the program and trigger recovery.
     */
    void apply(KEY key, W window, Iterable<IN> input, Collector<OUT> out) throws Exception;
}
```

It can be used like this:

Java | Scala

```java
DataStream<Tuple2<String, Long>> input = ...;

input
    .keyBy(<key selector>)
    .window(<window assigner>)
    .apply(new MyWindowFunction());
```

图7.16 apply全量聚合案例

```java
DataStream<Tuple2<String, Long>> input = ...;

input
    .keyBy(t -> t.f0)
    .timeWindow(Time.minutes(5))
    .process(new MyProcessWindowFunction());

/* ... */

public class MyProcessWindowFunction
        extends ProcessWindowFunction<Tuple2<String, Long>, String, String, TimeWindow> {

    @Override
    public void process(String key, Context context, Iterable<Tuple2<String, Long>> input, Collector<String> out) {
        long count = 0;
        for (Tuple2<String, Long> in: input) {
            count++;
        }
        out.collect("Window: " + context.window() + "count: " + count);
    }
}
```

图7.17 process全量聚合案例

第8章
Flink Time 详解

本章主要针对 Flink Time 中的 Event Time、Ingestion Time、Processing Time 以及 Watermark 进行详细讲解。

8.1 Time

Stream 数据中的 Time（时间）分为以下3种。

- Event Time：事件产生的时间，它通常由事件中的时间戳描述。
- Ingestion Time：事件进入 Flink 的时间。
- Processing Time：事件被处理时当前系统的时间。

这几种时间的对应关系如图 8.1 所示。

假设原始日志如下。

2019-01-10 10:00:01,134 INFO executor.Executor: Finished task in state 0.0

这条数据进入 Flink 的时间是 2019-01-10 20:00:00,102。

到达 Window 处理的时间为 2019-01-10 20:00:01,100。

如果我们想要统计每分钟内接口调用失败的错误日志个数，使用哪个时间才有意义？因为数据有可能出现延迟，所以使用数据进入 Flink 的时间或者 Window 处理的时间，其实是没有意义的，此时使用原始日志中的时间才是有意义的，那才是数据产生的时间。

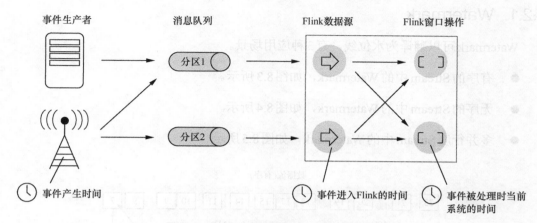

图8.1　Flink中的3种Time之间的关系

我们在Flink的Stream程序中处理数据时，默认使用的是哪个时间呢？如何修改呢？默认情况下，Flink在Stream程序中处理数据使用的时间是ProcessingTime，想要修改使用时间可以使用setStreamTimeCharacteristic()，代码如图8.2所示。

```
final StreamExecutionEnvironment env = StreamExecutionEnvironment.getExecutionEnvironment();

env.setStreamTimeCharacteristic(TimeCharacteristic.ProcessingTime);

// alternatively:
// env.setStreamTimeCharacteristic(TimeCharacteristic.IngestionTime);
// env.setStreamTimeCharacteristic(TimeCharacteristic.EventTime);
```

图8.2　代码设置使用哪种时间

8.2　Flink如何处理乱序数据

在使用EventTime处理Stream数据的时候会遇到数据乱序的问题，流处理从Event（事件）产生，流经Source，再到Operator，这中间需要一定的时间。虽然大部分情况下，传输到Operator的数据都是按照事件产生的时间顺序来的，但是也不排除由于网络延迟等原因而导致乱序的产生，特别是使用Kafka的时候，多个分区之间的数据无法保证有序。因此，在进行Window计算的时候，不能无限期地等下去，必须要有个机制来保证在特定的时间后，必须触发Window进行计算，这个特别的机制就是Watermark。Watermark是用于处理乱序事件的。

8.2.1 Watermark

Watermark 可以翻译为水位线，有3种应用场景。

- 有序的 Stream 中的 Watermark，如图8.3所示。
- 无序的 Stream 中的 Watermark，如图8.4所示。
- 多并行度 Stream 中的 Watermark，如图8.5所示。

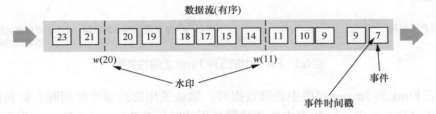

图8.3 有序的Stream中的Watermark

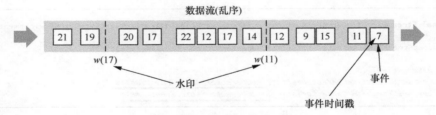

图8.4 无序的Stream中的Watermark

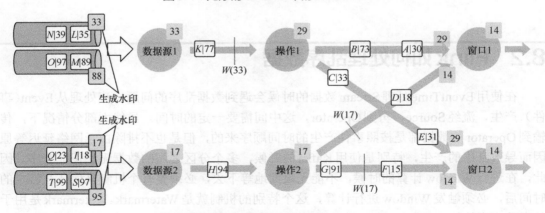

图8.5 多并行度Stream中的watermark

注意：在多并行度的情况下，Watermark 会有一个对齐机制，这个对齐机制会取所有 Channel 中最小的 Watermark，图 8.5 中的 14 和 29 这两个 Watermark 的最终取值为 14。

8.2.2 Watermark 的生成方式

通常情况下，在接收到 Source 的数据后，应该立刻生成 Watermark，但是也可以在应用简单的 Map 或者 Filter 操作后再生成 Watermark。

注意：如果指定多次 Watermark，后面指定的值会覆盖前面的值。

Watermark 的生成方式有两种。

1. With Periodic Watermarks

- 周期性地触发 Watermark 的生成和发送，默认是 100ms。
- 每隔 N 秒自动向流里注入一个 Watermark，时间间隔由 ExecutionConfig.setAutoWatermarkInterval 决定。每次调用 getCurrentWatermark 方法，如果得到的 Watermark 不为空并且比之前的大，就注入流中。
- 可以定义一个最大允许乱序的时间，这种比较常用。
- 实现 AssignerWithPeriodicWatermarks 接口。

2. With Punctuated Watermarks

- 基于某些事件触发 Watermark 的生成和发送。
- 基于事件向流里注入一个 Watermark，每一个元素都有机会判断是否生成一个 Watermark。如果得到的 Watermark 不为空并且比之前的大，就注入流中。
- 实现 AssignerWithPunctuatedWatermarks 接口。

第 1 种方式比较常用，所以在这里我们使用第 1 种方式进行分析。

参考官网文档中 With Periodic Watermarks 的使用方法，如图 8.6 所示。

图 8.6 所示代码中的 extractTimestamp 方法是从数据本身中提取 EventTime。getCurrentWatermar 方法是获取当前水位线，利用 currentMaxTimestamp - maxOutOfOrderness。maxOutOfOrderness 表示是允许数据的最大乱序时间。

```
/**
 * This generator generates watermarks assuming that elements arrive out of order,
 * but only to a certain degree. The latest elements for a certain timestamp t will arrive
 * at most n milliseconds after the earliest elements for timestamp t.
 */
public class BoundedOutOfOrdernessGenerator implements AssignerWithPeriodicWatermarks<MyEvent> {

    private final long maxOutOfOrderness = 3500; // 3.5 seconds

    private long currentMaxTimestamp;

    @Override
    public long extractTimestamp(MyEvent element, long previousElementTimestamp) {
        long timestamp = element.getCreationTime();
        currentMaxTimestamp = Math.max(timestamp, currentMaxTimestamp);
        return timestamp;
    }

    @Override
    public Watermark getCurrentWatermark() {
        // return the watermark as current highest timestamp minus the out-of-orderness bound
        return new Watermark(currentMaxTimestamp - maxOutOfOrderness);
    }
}
```

图8.6 With Periodic Watermarks的使用

在这里也需要实现接口AssignerWithPeriodicWatermarks，参考代码如图8.7所示。

```
final StreamExecutionEnvironment env = StreamExecutionEnvironment.getExecutionEnvironment();
env.setStreamTimeCharacteristic(TimeCharacteristic.EventTime);

DataStream<MyEvent> stream = env.readFile(
        myFormat, myFilePath, FileProcessingMode.PROCESS_CONTINUOUSLY, 100,
        FilePathFilter.createDefaultFilter(), typeInfo);

DataStream<MyEvent> withTimestampsAndWatermarks = stream
        .filter( event -> event.severity() == WARNING )
        .assignTimestampsAndWatermarks(new MyTimestampsAndWatermarks());

withTimestampsAndWatermarks
        .keyBy( (event) -> event.getGroup() )
        .timeWindow(Time.seconds(10))
        .reduce( (a, b) -> a.add(b) )
        .addSink(...);
```

图8.7 Watermark的使用

8.3 EventTime+Watermark解决乱序数据的案例详解

8.3.1 实现Watermark的相关代码

1. 程序说明

首先通过Socket模拟接收数据，然后使用map函数进行处理，接着调用assignTimestampsAndWatermarks方法抽取timestamp并生成Watermark，最后调用Window打印信息来验证Window被触发的时机。

8.3 EventTime+Watermark 解决乱序数据的案例详解

2. 代码实现

```java
package xuwei.tech.streaming.streamApiDemo;

import org.apache.Flink.api.common.functions.MapFunction;
import org.apache.Flink.api.java.tuple.Tuple;
import org.apache.Flink.api.java.tuple.Tuple2;
import org.apache.Flink.streaming.api.TimeCharacteristic;
import org.apache.Flink.streaming.api.DataStream.DataStream;
import org.apache.Flink.streaming.api.environment.StreamExecutionEnvironment;
import org.apache.Flink.streaming.api.functions.AssignerWithPeriodicWatermarks;
import org.apache.Flink.streaming.api.functions.windowing.WindowFunction;
import org.apache.Flink.streaming.api.watermark.Watermark;
import org.apache.Flink.streaming.api.windowing.assigners.TumblingEventTimeWindows;
import org.apache.Flink.streaming.api.windowing.time.Time;
import org.apache.Flink.streaming.api.windowing.windows.TimeWindow;
import org.apache.Flink.util.Collector;

import javax.annotation.Nullable;
import java.text.SimpleDateFormat;
import java.util.ArrayList;
import java.util.Collections;
import java.util.Iterator;
import java.util.List;


/**
 * Watermark案例
 * Created by xuwei.tech
 */
public class StreamingWindowWatermark {

    public static void main(String[] args) throws Exception {
        //定义Socket的端口号
        int port = 9000;
        //获取运行环境
        StreamExecutionEnvironment env = StreamExecutionEnvironment.getExecutionEnvironment();
        //设置使用EventTime,默认使用processtime
        env.setStreamTimeCharacteristic(TimeCharacteristic.EventTime);
```

```java
        //设置并行度为1,默认并行度是当前机器的CPU数量
        env.setParallelism(1);

        //连接Socket获取输入的数据
        DataStream<String> text = env.socketTextStream("hadoop100", port, "\n");

        //解析输入的数据
         DataStream<Tuple2<String, Long>> inputMap = text.map(new MapFunction<String
, Tuple2<String, Long>>() {
            @Override
            public Tuple2<String, Long> map(String value) throws Exception {
                String[] arr = value.split(",");
                return new Tuple2<>(arr[0], Long.parseLong(arr[1]));
            }
        });

        //抽取timestamp和生成Watermark
        DataStream<Tuple2<String, Long>> waterMarkStream = inputMap.assignTimestamps
AndWatermarks(new AssignerWithPeriodicWatermarks<Tuple2<String, Long>>() {

            Long currentMaxTimestamp = 0L;
            final Long maxOutOfOrderness = 10000L;// 最大允许的乱序时间是10s

            SimpleDateFormat sdf = new SimpleDateFormat("yyyy-MM-dd HH:mm:ss.SSS");
            /**
             * 定义生成Watermark的逻辑
             * 默认100ms被调用一次
             */
            @Nullable
            @Override
            public Watermark getCurrentWatermark() {
                return new Watermark(currentMaxTimestamp - maxOutOfOrderness);
            }

            //定义如何提取timestamp
            @Override
            public long extractTimestamp(Tuple2<String, Long> element, long previous
ElementTimestamp) {
                long timestamp = element.f1;
                currentMaxTimestamp = Math.max(timestamp, currentMaxTimestamp);
                 System.out.println("key:"+element.f0+",eventtime:["+element.f1+"|"+sdf.
format(element.f1)+"],currentMaxTimestamp:["+currentMaxTimestamp+"|"+
```

```java
                        sdf.format(currentMaxTimestamp)+"],watermark:["+getCurrentW
atermark().getTimestamp()+"|"+sdf.format(getCurrentWatermark().getTimestamp())+"]");
                return timestamp;
            }
        });

        //分组，聚合
        DataStream<String> window = waterMarkStream.keyBy(0)
                .window(TumblingEventTimeWindows.of(Time.seconds(3)))//按照消息的
EventTime分配窗口，和调用Time Window效果一样
                .apply(new WindowFunction<Tuple2<String, Long>, String, Tuple, TimeWindow>() {
                    /**
                     * 对Window内的数据进行排序，保证数据的顺序
                     * @param tuple
                     * @param window
                     * @param input
                     * @param out
                     * @throws Exception
                     */
                    @Override
                    public void apply(Tuple tuple, TimeWindow window, Iterable<Tuple2
<String, Long>> input, Collector<String> out) throws Exception {
                        String key = tuple.toString();
                        List<Long> arrarList = new ArrayList<Long>();
                        Iterator<Tuple2<String, Long>> it = input.iterator();
                        while (it.hasNext()) {
                            Tuple2<String, Long> next = it.next();
                            arrarList.add(next.f1);
                        }
                        Collections.sort(arrarList);
                        SimpleDateFormat sdf = new SimpleDateFormat("yyyy-MM-
dd HH:mm:ss.SSS");
                        String result = key + "," + arrarList.size() + "," + sdf.
format(arrarList.get(0)) + "," + sdf.format(arrarList.get(arrarList.size() - 1))
                                + "," + sdf.format(window.getStart()) + "," + sdf.
format(window.getEnd());
                        out.collect(result);
                    }
                });
        //测试，把结果打印到控制台即可
        window.print();

        //注意：因为Flink是懒加载的，所以必须调用execute方法，这样上面的代码才会执行
```

```
            env.execute("eventtime-watermark");
        }
```

3. 程序详解

（1）接收Socket数据。

（2）将每行数据按照逗号分隔，每行数据调用Map转换成Tuple<String,Long>类型。其中Tuple中的第1个元素代表具体的数据，第2个元素代表数据的EventTime。

（3）抽取Timestamp，生成Watermark，允许的最大乱序时间是10s，并打印（Key，EventTime，CurrentMaxTimestamp，Watermark）等信息。

（4）分组聚合，Window窗口大小为3s，输出（Key，窗口内元素个数，窗口内最初元素进入的时间，窗口内最后元素进入的时间，窗口自身开始时间，窗口自身结束时间）。

8.3.2 通过数据跟踪Watermark的时间

在这里重点查看Watermark和Timestamp的时间，通过数据的输出来确定Window的触发时机。

首先开启Socket，输入第一条数据。

```
[root@hadoop100 soft]# nc -l 9000
0001,1538359882000
```

输出的结果如图8.8所示。

```
key:0001,eventtime:[1538359882000|2018-10-01 10:11:22.000],currentMaxTimestamp:[1538359882000|2018-10-01 10:11:22.000],watermark:[1538359872000|2018-10-01 10:11:12.000]
```

图8.8　Watermark输出的结果

为了查看方便，我们把输入内容汇总到表格中，如表8.1所示。

表8.1　Watermark输出的结果

Key	EventTime	CurrentMaxTimeStamp	Watermark
0001	1538359882000	1538359882000	1538359872000
	2018-10-01 10:11:22.000	2018-10-01 10:11:22.000	2018-10-01 10:11:12.000

8.3 EventTime+Watermark 解决乱序数据的案例详解

此时，Watermark 的时间已经落后于 CurrentMaxTimeStamp10s 了，我们继续输入。

```
[root@hadoop100 soft]# nc -l 9000
0001,1538359882000
0001,1538359886000
```

此时，输出的结果如图 8.9 所示。

图 8.9 Watermark 输出的结果

我们再次汇总，如表 8.2 所示。

表 8.2 Watermark 输出的结果

Key	EventTime	CurrentMaxTimeStamp	Watermark
0001	1538359882000	1538359882000	1538359872000
	2018-10-01 10:11:22.000	2018-10-01 10:11:22.000	2018-10-01 10:11:12.000
0001	1538359886000	1538359886000	1538359876000
	2018-10-01 10:11:26.000	2018-10-01 10:11:26.000	2018-10-01 10:11:16.000

继续输入。

```
[root@hadoop100 soft]# nc -l 9000
0001,1538359882000
0001,1538359886000
0001,1538359892000
```

输出的结果内容如图 8.10 所示。

图 8.10 Watermark 输出的结果

汇总如表 8.3 所示。

表 8.3 Watermark 输出的结果

Key	EventTime	CurrentMaxTimeStamp	Watermark
0001	1538359882000	1538359882000	1538359872000
	2018-10-01 10:11:22.000	2018-10-01 10:11:22.000	2018-10-01 10:11:12.000

续表

Key	EventTime	CurrentMaxTimeStamp	Watermark
0001	1538359886000	1538359886000	1538359876000
	2018-10-01 10:11:26.000	2018-10-01 10:11:26.000	2018-10-01 10:11:16.000
0001	1538359892000	1538359892000	1538359882000
	2018-10-01 10:11:32.000	2018-10-01 10:11:32.000	2018-10-01 10:11:22.000

到这里，Window 仍然没有被触发，此时 Watermark 的时间已经等于第一条数据的 EventTime 了。那么 Window 到底什么时候被触发呢？我们再次输入。

```
[root@hadoop100 soft]# nc -l 9000
0001,1538359882000
0001,1538359886000
0001,1538359892000
0001,1538359893000
```

输出的结果如图 8.11 所示。

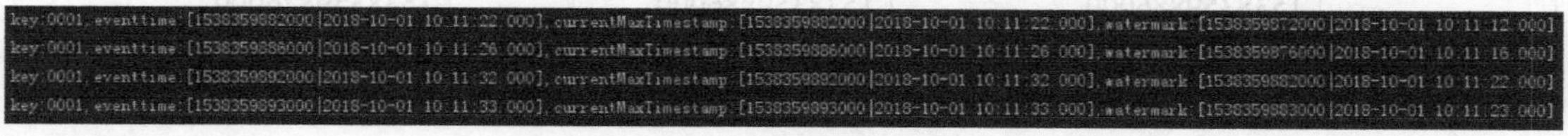

图 8.11 Watermark 输出的结果

汇总如表 8.4 所示。

表 8.4 Watermark 输出的结果

Key	EventTime	CurrentMaxTimeStamp	Watermark
0001	1538359882000	1538359882000	1538359872000
	2018-10-01 10:11:22.000	2018-10-01 10:11:22.000	2018-10-01 10:11:12.000
0001	1538359886000	1538359886000	1538359876000
	2018-10-01 10:11:26.000	2018-10-01 10:11:26.000	2018-10-01 10:11:16.000
0001	1538359892000	1538359892000	1538359882000
	2018-10-01 10:11:32.000	2018-10-01 10:11:32.000	2018-10-01 10:11:22.000
0001	1538359893000	1538359893000	1538359883000
	2018-10-01 10:11:33.000	2018-10-01 10:11:33.000	2018-10-01 10:11:23.000

Window 仍然没有触发，此时，我们的数据已经发到 2018-10-01 10:11:33.000 了。根据 EventTime 来算，距离最早的数据已经过去到达 11s 了，Window 还没有开始计算，到底什么时候会触发 Window 呢？

我们再次增加1s，输入。

```
[root@hadoop100 soft]# nc -l 9000
0001,1538359882000
0001,1538359886000
0001,1538359892000
0001,1538359893000
0001,1538359894000
```

输出的结果如图8.12所示。

图8.12　Watermark输出的结果

汇总如表8.5所示。

表8.5　Watermark输出的结果

Key	EventTime	CurrentMaxTimeStamp	Watermark	window_start_time	window_end_time
0001	1538359882000 2018-10-01 10:11:22.000	1538359882000 2018-10-01 10:11:22.000	1538359872000 2018-10-01 10:11:12.000		
0001	1538359886000 2018-10-01 10:11:26.000	1538359886000 2018-10-01 10:11:26.000	1538359876000 2018-10-01 10:11:16.000		
0001	1538359892000 2018-10-01 10:11:32.000	1538359892000 2018-10-01 10:11:32.000	1538359882000 2018-10-01 10:11:22.000		
0001	1538359893000 2018-10-01 10:11:33.000	1538359893000 2018-10-01 10:11:33.000	1538359883000 2018-10-01 10:11:23.000		
0001	1538359894000 2018-10-01 10:11:34.000	1538359894000 2018-10-01 10:11:34.000	1538359884000 2018-10-01 10:11:24.000	[10:11:21.000	10:11:24.000)

到这里，我们做一个说明。

Window 的触发机制是先按照自然时间对 Window 进行划分,如果 Window 的大小是 3s,那么 1min 内会把 Window 划分为如下的形式(左闭右开的区间)。

```
[00:00:00,00:00:03)
[00:00:03,00:00:06)
[00:00:06,00:00:09)
[00:00:09,00:00:12)
[00:00:12,00:00:15)
[00:00:15,00:00:18)
[00:00:18,00:00:21)
[00:00:21,00:00:24)
[00:00:24,00:00:27)
[00:00:27,00:00:30)
[00:00:30,00:00:33)
[00:00:33,00:00:36)
[00:00:36,00:00:39)
[00:00:39,00:00:42)
[00:00:42,00:00:45)
[00:00:45,00:00:48)
[00:00:48,00:00:51)
[00:00:51,00:00:54)
[00:00:54,00:00:57)
[00:00:57,00:01:00)
...
```

Window 的设定无关数据本身,而是系统定义好的。

输入数据时,根据自身的 EventTime 将其划分到不同的 Window 中。如果 Window 中有数据,则当 Watermark 时间 ≥ EventTime 时,就符合 Window 触发的条件,但最终是否决定 Window 触发,还是由数据本身的 EventTime 所属的 Window 中的 window_end_time 决定。

上面的测试中,最后一条数据到达后,其水位线已经升至 10:11:24,正好是最早的一条记录所在 Window 的 window_end_time,所以 Window 就被触发了。

为了验证 Window 的触发机制,我们继续输入数据。

```
[root@hadoop100 soft]# nc -l 9000
0001,1538359882000
0001,1538359886000
0001,1538359892000
0001,1538359893000
0001,1538359894000
0001,1538359896000
```

输出的结果如图 8.13 所示。

```
key:0001,eventtime:[1538359882000|2018-10-01 10:11:22.000],currentMaxTimestamp:[1538359882000|2018-10-01 10:11:22.000],watermark:[1538359872000|2018-10-01 10:11:12.000]
key:0001,eventtime:[1538359886000|2018-10-01 10:11:26.000],currentMaxTimestamp:[1538359886000|2018-10-01 10:11:26.000],watermark:[1538359876000|2018-10-01 10:11:16.000]
key:0001,eventtime:[1538359892000|2018-10-01 10:11:32.000],currentMaxTimestamp:[1538359892000|2018-10-01 10:11:32.000],watermark:[1538359882000|2018-10-01 10:11:22.000]
key:0001,eventtime:[1538359893000|2018-10-01 10:11:33.000],currentMaxTimestamp:[1538359893000|2018-10-01 10:11:33.000],watermark:[1538359883000|2018-10-01 10:11:23.000]
key:0001,eventtime:[1538359894000|2018-10-01 10:11:34.000],currentMaxTimestamp:[1538359894000|2018-10-01 10:11:34.000],watermark:[1538359884000|2018-10-01 10:11:24.000]
(0001),1,2018-10-01 10:11:22.000,2018-10-01 10:11:22.000,2018-10-01 10:11:21.000,2018-10-01 10:11:24.000
key:0001,eventtime:[1538359896000|2018-10-01 10:11:36.000],currentMaxTimestamp:[1538359896000|2018-10-01 10:11:36.000],watermark:[1538359886000|2018-10-01 10:11:26.000]
```

图 8.13　Watermark 输出的结果

汇总如表 8.6 所示。

表 8.6　Watermark 输出的结果

Key	EventTime	CurrentMaxTimeStamp	Watermark	window_start_time	window_end_time
0001	1538359882000	1538359882000	1538359872000		
	2018-10-01 10:11:22.000	2018-10-01 10:11:22.000	2018-10-01 10:11:12.000		
0001	1538359886000	1538359886000	1538359876000		
	2018-10-01 10:11:26.000	2018-10-01 10:11:26.000	2018-10-01 10:11:16.000		
0001	1538359892000	1538359892000	1538359882000		
	2018-10-01 10:11:32.000	2018-10-01 10:11:32.000	2018-10-01 10:11:22.000		
0001	1538359893000	1538359893000	1538359883000		
	2018-10-01 10:11:33.000	2018-10-01 10:11:33.000	2018-10-01 10:11:23.000		
0001	1538359894000	1538359894000	1538359884000		
	2018-10-01 10:11:34.000	2018-10-01 10:11:34.000	2018-10-01 10:11:24.000	[10:11:21.000	10:11:24.000)
0001	1538359896000	1538359896000	1538359886000		
	2018-10-01 10:11:36.000	2018-10-01 10:11:36.000	2018-10-01 10:11:26.000		

此时，Watermark 时间虽然已经达到了第二条数据的时间，但是由于其没有达到第二条数据所在 Window 的结束时间，因此 Window 并没有被触发。第二条数据所在的 Window 时

间区间如下。

```
[00:00:24,00:00:27)
```

也就是说，必须输入一个10:11:27的数据，第二条数据所在的Window才会被触发。接下来继续输入数据。

```
[root@hadoop100 soft]# nc -l 9000
0001,1538359882000
0001,1538359886000
0001,1538359892000
0001,1538359893000
0001,1538359894000
0001,1538359896000
0001,1538359897000
```

输出的结果如图8.14所示。

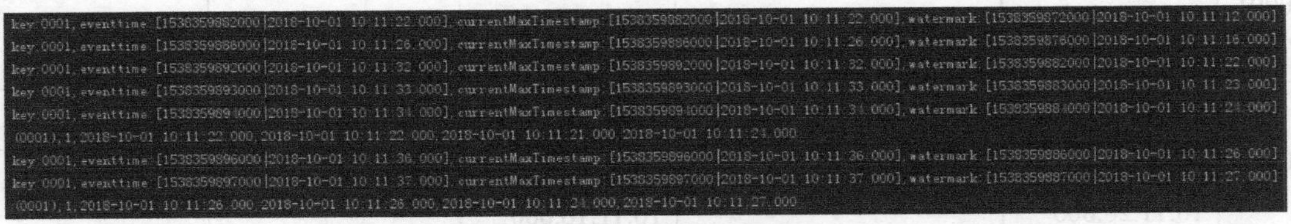

图8.14 Watermark输出的结果

汇总如表8.7所示。

表8.7 Watermark输出的结果

Key	EventTime	CurrentMaxTimeStamp	WaterMark	window_start_time	window_end_time
0001	1538359882000	1538359882000	1538359872000		
	2018-10-01 10:11:22.000	2018-10-01 10:11:22.000	2018-10-01 10:11:12.000		
0001	1538359886000	1538359886000	1538359876000		
	2018-10-01 10:11:26.000	2018-10-01 10:11:26.000	2018-10-01 10:11:16.000		
0001	1538359892000	1538359892000	1538359882000		
	2018-10-01 10:11:32.000	2018-10-01 10:11:32.000	2018-10-01 10:11:22.000		

Key	EventTime	CurrentMaxTimeStamp	WaterMark	window_start_time	window_end_time
0001	1538359893000 2018-10-01 10:11:33.000	1538359893000 2018-10-01 10:11:33.000	1538359883000 2018-10-01 10:11:23.000		
0001	1538359894000 2018-10-01 10:11:34.000	1538359894000 2018-10-01 10:11:34.000	1538359884000 2018-10-01 10:11:24.000	[10:11:21.000	10:11:24.000)
0001	1538359896000 2018-10-01 10:11:36.000	1538359896000 2018-10-01 10:11:36.000	1538359886000 2018-10-01 10:11:26.000		
0001	1538359897000 2018-10-01 10:11:37.000	1538359897000 2018-10-01 10:11:37.000	1538359887000 2018-10-01 10:11:27.000	[10:11:24.000	10:11:27.000)

此时，我们已经看到，Window 的触发要符合以下几个条件。

- Watermark 时间≥window_end_time。
- 在 [window_start_time, window_end_time) 中有数据存在（注意是左闭右开的区间）。

同时满足了以上 2 个条件，Window 才会触发。

8.3.3 利用 Watermark+Window 处理乱序数据

在上面的测试中，数据都是按照时间顺序递增的。现在，输入一些乱序的（Late）数据，看一看 Watermark 结合 Window 机制是如何处理乱序的。

输入两行数据。

```
[root@hadoop100 soft]# nc -l 9000
0001,1538359882000
0001,1538359886000
0001,1538359892000
0001,1538359893000
0001,1538359894000
0001,1538359896000
0001,1538359897000
```

```
0001,1538359899000
0001,1538359891000
```

输出的结果如图8.15所示。

图8.15 Watermark输出的结果

汇总如表8.8所示。

表8.8 Watermark输出的结果

Key	EventTime	CurrentMaxTimeStamp	WaterMark	window_start_time	window_end_time
0001	1538359882000 2018-10-01 10:11:22.000	2018-10-01 10:11:22.000	1538359872000 2018-10-01 10:11:12.000		
0001	1538359886000 2018-10-01 10:11:26.000	2018-10-01 10:11:26.000	1538359876000 2018-10-01 10:11:16.000		
0001	1538359892000 2018-10-01 10:11:32.000	2018-10-01 10:11:32.000	1538359882000 2018-10-01 10:11:22.000		
0001	1538359893000 2018-10-01 10:11:33.000	2018-10-01 10:11:33.000	1538359883000 2018-10-01 10:11:23.000		
0001	1538359894000 2018-10-01 10:11:34.000	2018-10-01 10:11:34.000	1538359884000 2018-10-01 10:11:24.000	[10:11:21.000	10:11:24.000)
0001	1538359896000 2018-10-01 10:11:36.000	2018-10-01 10:11:36.000	1538359886000 2018-10-01 10:11:26.000		

续表

Key	EventTime	CurrentMaxTimeStamp	WaterMark	window_start_time	window_end_time
0001	1538359897000 2018-10-01 10:11:37.000	1538359897000 2018-10-01 10:11:37.000	1538359887000 2018-10-01 10:11:27.000	[10:11:24.000	10:11:27.000)
0001	1538359899000 2018-10-01 10:11:39.000	1538359899000 2018-10-01 10:11:39.000	1538359889000 2018-10-01 10:11:29.000		
0001	1538359891000 2018-10-01 10:11:31.000	1538359899000 2018-10-01 10:11:39.000	1538359889000 2018-10-01 10:11:29.000		

可以看到，虽然我们输入了一个10:11:31的数据，但是currentMaxTimestamp和Watermark都没变。此时，按照上面提到的公式。

- Watermark时间≥window_end_time。

- 在[window_start_time,window_end_time)中有数据存在。

Watermark时间（10:11:29）< window_end_time(10:11:33)，因此不能触发Window。

如果再次输入一条10:11:43的数据，此时Watermark时间会升高到10:11:33，这时的Window一定会触发。我们试一试，继续输入内容。

```
[root@hadoop100 soft]# nc -l 9000
0001,1538359882000
0001,1538359886000
0001,1538359892000
0001,1538359893000
0001,1538359894000
0001,1538359896000
0001,1538359897000
0001,1538359899000
0001,1538359891000
0001,1538359903000
```

输出的结果如图8.16所示。

图8.16 Watermark输出的结果

汇总如表8.9所示。

表8.9 Watermark输出的结果

Key	EventTime	CurrentMaxTimeStamp	Watermark	window_start_time	window_end_time
0001	1538359882000 2018-10-01 10:11:22.000	1538359882000 2018-10-01 10:11:22.000	1538359872000 2018-10-01 10:11:12.000		
0001	1538359886000 2018-10-01 10:11:26.000	1538359886000 2018-10-01 10:11:26.000	1538359876000 2018-10-01 10:11:16.000		
0001	1538359892000 2018-10-01 10:11:32.000	1538359892000 2018-10-01 10:11:32.000	1538359882000 2018-10-01 10:11:22.000		
0001	1538359893000 2018-10-01 10:11:33.000	1538359893000 2018-10-01 10:11:33.000	1538359883000 2018-10-01 10:11:23.000		
0001	1538359894000 2018-10-01 10:11:34.000	1538359894000 2018-10-01 10:11:34.000	1538359884000 2018-10-01 10:11:24.000	[10:11:21.000	10:11:24.000)
0001	1538359896000 2018-10-01 10:11:36.000	1538359896000 2018-10-01 10:11:36.000	1538359886000 2018-10-01 10:11:26.000		
0001	1538359897000 2018-10-01 10:11:37.000	1538359897000 2018-10-01 10:11:37.000	1538359887000 2018-10-01 10:11:27.000	[10:11:24.000	10:11:27.000)

续表

Key	EventTime	CurrentMaxTimeStamp	Watermark	window_start_time	window_end_time
0001	1538359899000 2018-10-01 10:11:39.000	1538359899000 2018-10-01 10:11:39.000	1538359889000 2018-10-01 10:11:29.000		
0001	1538359891000 2018-10-01 10:11:31.000	1538359899000 2018-10-01 10:11:39.000	1538359889000 2018-10-01 10:11:29.000		
0001	1538359903000 2018-10-01 10:11:43.000	1538359903000 2018-10-01 10:11:43.000	1538359893000 2018-10-01 10:11:33.000	[10:11:30.000	10:11:33.000)

这里可以看到，窗口中有2个数据：10:11:31和10:11:32，但是没有10:11:33的数据，因为窗口是一个前闭后开的区间，10:11:33的数据是属于[10:11:33,10:11:36)这个窗口的。

上面的结果已经表明，Flink可以通过Watermark机制结合Window的操作来处理一定范围内的乱序数据。那么对于Late Element（延迟数据），Flink是怎么处理的呢？

8.3.4　Late Element的处理方式

针对延迟数据，Flink有3种处理方案。

1．丢弃（默认）

输入一个乱序的（其实只要EventTime < Watermark时间）数据来测试。

输入两行内容。

```
[root@hadoop100 soft]# nc -l 9000
0001,1538359890000
0001,1538359903000
```

输出的结果如图8.17所示。

```
key 0001,eventtime:[1538359890000|2018-10-01 10:11:30.000],currentMaxTimestamp:[1538359890000|2018-10-01 10:11:30.000],watermark:[1538359880000|2018-10-01 10:11:20.000]
key 0001,eventtime:[1538359903000|2018-10-01 10:11:43.000],currentMaxTimestamp:[1538359903000|2018-10-01 10:11:43.000],watermark:[1538359893000|2018-10-01 10:11:33.000]
(0001),1,2018-10-01 10:11:30.000,2018-10-01 10:11:30.000,2018-10-01 10:11:30.000,2018-10-01 10:11:33.000
```

图 8.17 Watermark 输出的结果

汇总如表 8.10 所示。

表 8.10 Watermark 输出的结果

Key	EventTime	CurrentMaxTimeStamp	Watermark	window_start_time	window_end_time
0001	1538359890000	1538359890000	1538359880000		
	2018-10-01 10:11:30.000	2018-10-01 10:11:30.000	2018-10-01 10:11:20.000		
0001	1538359903000	1538359903000	1538359893000		
	2018-10-01 10:11:43.000	2018-10-01 10:11:43.000	2018-10-01 10:11:33.000	[10:11:30.000	10:11:33.000)

注意：此时 Watermark 是 2018-10-01 10:11:33.000。

下面再输入几个 EventTime<Watermark 的时间。

输入 3 行内容。

```
[root@hadoop100 soft]# nc -l 9000
0001,1538359890000
0001,1538359903000
0001,1538359890000
0001,1538359891000
0001,1538359892000
```

输出的结果如图 8.18 所示。

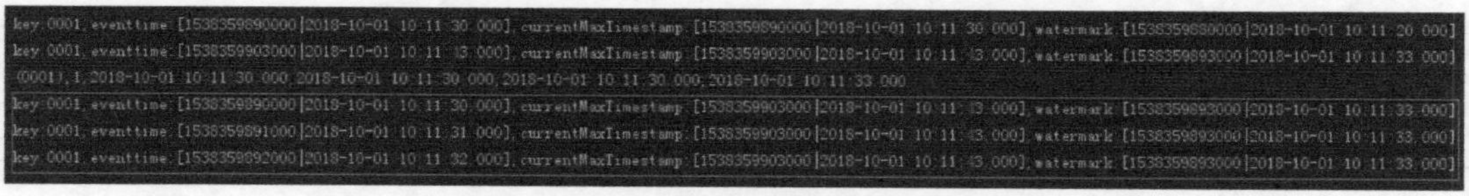

```
key 0001,eventtime:[1538359890000|2018-10-01 10:11:30.000],currentMaxTimestamp:[1538359890000|2018-10-01 10:11:30.000],watermark:[1538359880000|2018-10-01 10:11:20.000]
key 0001,eventtime:[1538359903000|2018-10-01 10:11:43.000],currentMaxTimestamp:[1538359903000|2018-10-01 10:11:43.000],watermark:[1538359893000|2018-10-01 10:11:33.000]
(0001),1,2018-10-01 10:11:30.000,2018-10-01 10:11:30.000,2018-10-01 10:11:30.000,2018-10-01 10:11:33.000
key 0001,eventtime:[1538359890000|2018-10-01 10:11:30.000],currentMaxTimestamp:[1538359903000|2018-10-01 10:11:43.000],watermark:[1538359893000|2018-10-01 10:11:33.000]
key 0001,eventtime:[1538359891000|2018-10-01 10:11:31.000],currentMaxTimestamp:[1538359903000|2018-10-01 10:11:43.000],watermark:[1538359893000|2018-10-01 10:11:33.000]
key 0001,eventtime:[1538359892000|2018-10-01 10:11:32.000],currentMaxTimestamp:[1538359903000|2018-10-01 10:11:43.000],watermark:[1538359893000|2018-10-01 10:11:33.000]
```

图 8.18 Watermark 输出的结果

注意：此时并没有触发 Window。因为输入的数据所在的窗口已经执行过了，Flink 默认对这些延迟的数据的处理方案就是丢弃。

2. allowedLateness 指定允许数据延迟的时间

在某些情况下，我们希望为延迟的数据提供一个宽容的时间。

Flink 提供了 allowedLateness 方法，它可以实现对延迟的数据设置一个延迟时间，在指定延迟时间内到达的数据可以触发 Window。

修改代码如图 8.19 所示。

```
//分组，聚合
DataStream<String> window = waterMarkStream.keyBy(0)
                .window(TumblingEventTimeWindows.of(Time.seconds(3)))//按照消息的EventTime分配窗口，和调用
                .allowedLateness(Time.seconds(2))//允许数据迟到2秒
                .apply(new WindowFunction<Tuple2<String, Long>, String, Tuple, TimeWindow>() {
                    /**
                     * 对window内的数据进行排序，保证数据的顺序
                     * @param
```

图 8.19　修改代码

下面来验证一下，输入 2 行内容。

```
[root@hadoop100 soft]# nc -l 9000
0001,1538359890000
0001,1538359903000
```

输出的结果如图 8.20 所示。

```
key:0001,eventtime:[1538359890000|2018-10-01 10:11:30.000],currentMaxTimestamp:[1538359890000|2018-10-01 10:11:30.000],watermark:[1538359880000|2018-10-01 10:11:20.000]
key:0001,eventtime:[1538359903000|2018-10-01 10:11:43.000],currentMaxTimestamp:[1538359903000|2018-10-01 10:11:43.000],watermark:[1538359893000|2018-10-01 10:11:33.000]
(0001),1,2018-10-01 10:11:30.000,2018-10-01 10:11:30.000,2018-10-01 10:11:30.000,2018-10-01 10:11:33.000
```

图 8.20　Watermark 输出的结果

这里正常触发了 Window。

汇总如表 8.11 所示。

表 8.11　Watermark 输出的结果

Key	EventTime	CurrentMaxTimeStamp	WaterMark	window_start_time	window_end_time
0001	1538359890000 2018-10-01 10:11:30.000	1538359890000 2018-10-01 10:11:30.000	1538359880000 2018-10-01 10:11:20.000		

续表

Key	EventTime	CurrentMaxTimeStamp	WaterMark	window_start_time	window_end_time
0001	1538359903000	1538359903000	1538359893000		
	2018-10-01 10:11:43.000	2018-10-01 10:11:43.000	2018-10-01 10:11:33.000	[10:11:33.000	10:11:33.000)

此时Watermark是2018-10-01 10:11:33.000，那么现在输入几条EventTime<Watermark的数据来验证一下效果，输入3行内容。

```
[root@hadoop100 soft]# nc -l 9000
0001,1538359890000
0001,1538359903000
0001,1538359890000
0001,1538359891000
0001,1538359892000
```

输出的结果如图8.21所示。

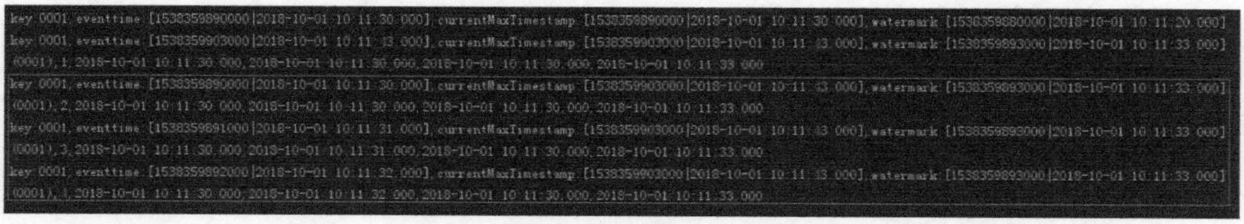

图8.21 Watermark输出的结果

在这里看到每条数据都触发了Window执行。

汇总如表8.12所示。

表8.12 Watermark输出的结果

Key	EventTime	CurrentMaxTimeStamp	WaterMark	window_start_time	window_end_time
0001	1538359890000	1538359890000	1538359880000		
	2018-10-01 10:11:30.000	2018-10-01 10:11:30.000	2018-10-01 10:11:20.000		
0001	1538359903000	1538359903000	1538359893000		
	2018-10-01 10:11:43.000	2018-10-01 10:11:43.000	2018-10-01 10:11:33.000	[10:11:33.000	10:11:33.000)

8.3 EventTime+Watermark 解决乱序数据的案例详解

续表

Key	EventTime	CurrentMaxTimeStamp	WaterMark	window_start_time	window_end_time
0001	1538359890000	1538359903000	1538359893000		
	2018-10-01 10:11:30.000	2018-10-01 10:11:43.000	2018-10-01 10:11:33.000	[10:11:33.000	10:11:33.000)
0001	1538359891000	1538359903000	1538359893000		
	2018-10-01 10:11:31.000	2018-10-01 10:11:43.000	2018-10-01 10:11:33.000	[10:11:33.000	10:11:33.000)
0001	1538359892000	1538359903000	1538359893000		
	2018-10-01 10:11:32.000	2018-10-01 10:11:43.000	2018-10-01 10:11:33.000	[10:11:33.000	10:11:33.000)

我们再输入1条数据，把Watermark调整到10:11:34。

```
[root@hadoop100 soft]# nc -l 9000
0001,1538359890000
0001,1538359903000
0001,1538359890000
0001,1538359891000
0001,1538359892000
0001,1538359904000
```

输出的结果如图8.22所示。

图8.22 Watermark输出的结果

汇总如表8.13所示。

表8.13 Watermark输出的结果

Key	EventTime	CurrentMaxTimeStamp	WaterMark	window_start_time	window_end_time
0001	1538359890000	1538359890000	1538359880000		
	2018-10-01 10:11:30.000	2018-10-01 10:11:30.000	2018-10-01 10:11:20.000		
0001	1538359903000	1538359903000	1538359893000		
	2018-10-01 10:11:43.000	2018-10-01 10:11:43.000	2018-10-01 10:11:33.000	[10:11:33.000	10:11:33.000)
0001	1538359890000	1538359903000	1538359893000		
	2018-10-01 10:11:30.000	2018-10-01 10:11:43.000	2018-10-01 10:11:33.000	[10:11:33.000	10:11:33.000)
0001	1538359891000	1538359903000	1538359893000		
	2018-10-01 10:11:31.000	2018-10-01 10:11:43.000	2018-10-01 10:11:33.000	[10:11:33.000	10:11:33.000)
0001	1538359892000	1538359903000	1538359893000		
	2018-10-01 10:11:32.000	2018-10-01 10:11:43.000	2018-10-01 10:11:33.000	[10:11:33.000	10:11:33.000)
0001	1538359904000	1538359904000	1538359894000		
	2018-10-01 10:11:44.000	2018-10-01 10:11:44.000	2018-10-01 10:11:34.000		

此时,把Watermark上升到了10:11:34,再输入几条EventTime<Watermark的数据来验证一下效果,输入3行内容。

```
[root@hadoop100 soft]# nc -l 9000
0001,1538359890000
0001,1538359903000
0001,1538359890000
0001,1538359891000
0001,1538359892000
0001,1538359904000
0001,1538359890000
0001,1538359891000
0001,1538359892000
```

输出的结果如图8.23所示。

图8.23　Watermark输出的结果

发现输入的3行数据都触发了Window的执行。

我们再输入1行数据，把Watermark调整到10:11:35。

```
[root@hadoop100 soft]# nc -l 9000
0001,1538359890000
0001,1538359903000
0001,1538359890000
0001,1538359891000
0001,1538359892000
0001,1538359904000
0001,1538359890000
0001,1538359891000
0001,1538359892000
0001,1538359905000
```

输出的结果如图8.24所示。

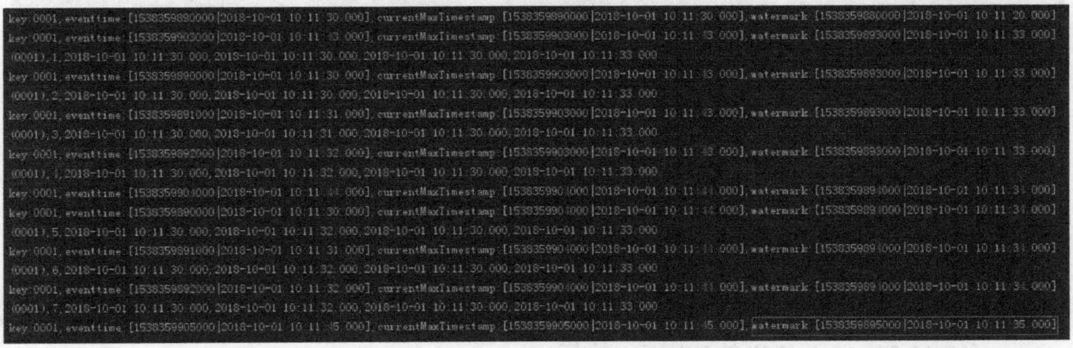

图8.24　Watermark输出的结果

此时，Watermark 上升到了 10:11:35。

再输入几条 EventTime<Watermark 的数据来验证一下效果，输入3条数据。

```
[root@hadoop100 soft]# nc -l 9000
0001,1538359890000
0001,1538359903000
0001,1538359890000
0001,1538359891000
0001,1538359892000
0001,1538359904000
0001,1538359890000
0001,1538359891000
0001,1538359892000
0001,1538359905000
0001,1538359890000
0001,1538359891000
0001,1538359892000
```

输出的结果如图8.25所示。

图8.25　Watermark输出的结果

发现这几条数据都没有触发 Window。

分析如下。

- 当 Watermark 等于 10:11:33 的时候，正好是 window_end_time，所以会触发 [10:11:30~10:11:33) 的 Window 执行。

- 当窗口执行过后，我们输入 [10:11:30~10:11:33) 这个 Window 内的数据，会发现 Window 是可以被触发的。

- 当 Watermark 提升到 10:11:34 的时候，输入 [10:11:30~10:11:33) 这个 Window 内的数据，会发现 Window 也是可以被触发的。

- 当 Watermark 提升到 10:11:35 的时候，输入 [10:11:30~10:11:33) 这个 Window 内的数据，会发现 Window 不会被触发。

由于在前面设置了 allowedLateness(Time.seconds(2))，因此可以允许延迟在 2s 内的数据继续触发 Window 执行。当 Watermark 为 10:11:34 时可以触发 Window，但是当 Watermark 为 10:11:35 时就不行了。

总结如下。

- 对于此窗口而言，允许 2s 的延迟时间，即第一次触发是在 Watermark ≥ window_end_time 时。

- 第二次（或多次）触发的条件是 Watermark < window_end_time + allowedLateness，这个窗口有延迟数据到达时。

分析如下。

- 当 Watermark 等于 10:11:34 的时候，输入 EventTime 为 10:11:30、10:11:31、10:11:32 的数据，是可以触发 Window 的，因为这些数据的 window_end_time 都是 10:11:33，也就是 10:11:34<10:11:33+2 为 true。

- 但是当 Watermark 等于 10:11:35 的时候，再输入 EventTime 为 10:11:30、10:11:31、10:11:32 的数据，这些数据的 window_end_time 都是 10:11:33，此时，10:11:35<10:11:33+2 为 false，因此最终这些数据延迟的时间太久，就不会再触发 Window 的执行操作了。

3. sideOutputLateData 收集延迟数据

通过 sideOutputLateData 函数可以把延迟数据统一收集、统一存储，方便后期排查问题。

需要先修改代码，如图 8.26 所示。

```java
//保存被丢弃的数据
OutputTag<Tuple2<String, Long>> outputTag = new OutputTag<Tuple2<String, Long>>( id: "late-data"){};
//注意，由于getSideOutput方法是SingleOutputStreamOperator子类中的特有方法，所以这里的类型，不能使用它的父类dataStream。
SingleOutputStreamOperator<String> window = waterMarkStream.keyBy( ...fields: 0)
    .window(TumblingEventTimeWindows.of(Time.seconds(3)))//按照消息的EventTime分配窗口，和调用TimeWindow效果一样
    //.allowedLateness(Time.seconds(2))//允许数据迟到2秒
    .sideOutputLateData(outputTag)
    .apply(new WindowFunction<Tuple2<String, Long>, String, Tuple, TimeWindow>() {
        /**
         * 对window内的数据进行排序，保证数据的顺序
         * @param
         * @param
         * @param
         * @param
         * @throws Exception
         */
        @Override
        public void apply(Tuple tuple, TimeWindow window, Iterable<Tuple2<String, Long>> input, Collector<String
            String key = tuple.toString();
            List<Long> arrarList = new ArrayList<Long>();
            Iterator<Tuple2<String, Long>> it = input.iterator();
            while (it.hasNext()) {
                Tuple2<String, Long> next = it.next();
                arrarList.add(next.f1);
            }
            Collections.sort(arrarList);
            SimpleDateFormat sdf = new SimpleDateFormat( pattern: "yyyy-MM-dd HH:mm:ss.SSS");
            String result = key + "," + arrarList.size() + "," + sdf.format(arrarList.get(0)) + "," + sdf.format
                + "," + sdf.format(window.getStart()) + "," + sdf.format(window.getEnd());
            out.collect(result);
        }
    });
//把迟到的数据暂时打印到控制台，实际中可以保存到其他存储介质中
DataStream<Tuple2<String, Long>> sideOutput = window.getSideOutput(outputTag);
sideOutput.print();
//测试-把结果打印到控制台即可
window.print();

//注意，因为flink是懒加载的，所以必须调用execute方法，上面的代码才会执行
env.execute( jobName: "eventtime-watermark");
```

图8.26 修改代码

我们输入两行数据来验证一下。

```
[root@hadoop100 soft]# nc -l 9000
0001,1538359890000
0001,1538359903000
```

输出的结果如图8.27所示。

```
key:0001,eventtime:[1538359890000|2018-10-01 10:11:30.000],currentMaxTimestamp:[1538359890000|2018-10-01 10:11:30.000],watermark:[1538359888000|2018-10-01 10:11:20.000]
key:0001,eventtime:[1538359903000|2018-10-01 10:11:43.000],currentMaxTimestamp:[1538359903000|2018-10-01 10:11:43.000],watermark:[1538359893000|2018-10-01 10:11:33.000]
(0001),1,2018-10-01 10:11:30.000,2018-10-01 10:11:30.000,2018-10-01 10:11:30.000,2018-10-01 10:11:33.000
```

图8.27 Watermark输出结果

此时，Window被触发执行了，Watermark是10:11:33。

下面再输入3行EventTime<Watermark的数据来进行测试。

```
[root@hadoop100 soft]# nc -l 9000
0001,1538359890000
0001,1538359903000
0001,1538359890000
0001,1538359891000
0001,1538359892000
```

输出的结果如图8.28所示。

```
key:0001,eventtime:[1538359890000|2018-10-01 10:11:30.000],currentMaxTimestamp:[1538359890000|2018-10-01 10:11:30.000],watermark:[1538359880000|2018-10-01 10:11:20.000]
key:0001,eventtime:[1538359903000|2018-10-01 10:11:43.000],currentMaxTimestamp:[1538359903000|2018-10-01 10:11:43.000],watermark:[1538359893000|2018-10-01 10:11:33.000]
(0001),1,2018-10-01 10:11:30.000,2018-10-01 10:11:30.000,2018-10-01 10:11:33:000
key:0001,eventtime:[1538359890000|2018-10-01 10:11:30.000],currentMaxTimestamp:[1538359903000|2018-10-01 10:11:43.000],watermark:[1538359893000|2018-10-01 10:11:33.000]
(0001,1538359890000)
key:0001,eventtime:[1538359891000|2018-10-01 10:11:31.000],currentMaxTimestamp:[1538359903000|2018-10-01 10:11:43.000],watermark:[1538359893000|2018-10-01 10:11:33.000]
(0001,1538359891000)
key:0001,eventtime:[1538359892000|2018-10-01 10:11:32.000],currentMaxTimestamp:[1538359903000|2018-10-01 10:11:43.000],watermark:[1538359893000|2018-10-01 10:11:33.000]
(0001,1538359892000)
```

图8.28　Watermark输出的结果

此时，这几条延迟的数据都通过sideOutputLateData保存到了outputTag中。

8.3.5　在多并行度下的Watermark应用

前面的代码将并行度设置为1。

```
env.setParallelism(1);
```

如果这里不设置的话，代码在IDEA中运行的时候会通过默认读取本机的CPU数量来设置并行度。把并行度注释掉的代码如图8.29所示。

```
//设置并行度为1,默认并行度是当前机器的CPU数量
//env.setParallelism(1);
```

图8.29　注释掉并行度的代码

然后在输出内容前面加上线程ID信息，如图8.30所示。

```
//定义如何提取timestamp
@Override
public long extractTimestamp(Tuple2<String, Long> element, long previousElementTimestamp) {
    long timestamp = element.f1;
    currentMaxTimestamp = Math.max(timestamp, currentMaxTimestamp);
    long id = Thread.currentThread().getId();
    System.out.println("currentThreadId:"+id+",key:"+element.f0+",eventtime:["+element.f1+"|"+sdf.format(ele
            sdf.format(currentMaxTimestamp)+"],watermark:["+getCurrentWatermark().getTimestamp()+"|"+sdf.for
    return timestamp;
}
```

图8.30　加上线程ID信息

输入如下7行内容。

```
[root@hadoop100 soft]# nc -l 9000
0001,1538359882000
0001,1538359886000
0001,1538359892000
0001,1538359893000
0001,1538359894000
0001,1538359896000
0001,1538359897000
```

输出的结果如图8.31所示。

图8.31　Watermark输出的结果

此时Window没有被触发，因为这7条数据都是被不同的线程处理的，每个线程都有一个Watermark。

在多并行度的情况下，Watermark对齐机制会取所有Channel中最小的Watermark。但是现在默认有8个并行度，这7条数据都被不同的线程所处理，到现在还没获取最小的Watermark，因此Window无法被触发执行，如图8.32所示。

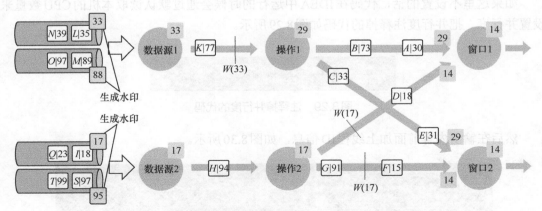

图8.32　多并行度下的Watermark

下面把代码中的并行度调整为2，如图8.33所示。

```
//设置并行度为1,默认并行度是当前机器的CPU数量
env.setParallelism(2);
```

图8.33 修改并行度代码

输入如下内容。

```
[root@hadoop100 soft]# nc -l 9000
0001,1538359890000
0001,1538359903000
0001,1538359908000
```

输出的结果如图8.34所示。

```
currentThreadId:44,key:0001,eventtime:[1538359890000|2018-10-01 10:11:30.000],currentMaxTimestamp:[1538359890000|2018-10-01 10:11:30.000],watermark:[1538359880000|2018-10-01 10:11:20.000]
currentThreadId:42,key:0001,eventtime:[1538359903000|2018-10-01 10:11:43.000],currentMaxTimestamp:[1538359903000|2018-10-01 10:11:43.000],watermark:[1538359893000|2018-10-01 10:11:33.000]
currentThreadId:44,key:0001,eventtime:[1538359908000|2018-10-01 10:11:48.000],currentMaxTimestamp:[1538359908000|2018-10-01 10:11:48.000],watermark:[1538359898000|2018-10-01 10:11:38.000]
1> (0001,1,2018-10-01 10:11:30.000,2018-10-01 10:11:30.000,2018-10-01 10:11:30.000,2018-10-01 10:11:33.000)
```

图8.34 Watermark输出的结果

此时会发现,当第3条数据输入完成以后,[10:11:30,10:11:33)这个Window被触发了。输入前两条数据之后,获取的最小Watermark是10:11:20,这时对应的Window中没有数据。输入第3条数据之后,获取的最小Watermark是10:11:33,这时对应的窗口就是[10:11:30,10:11:33),所以就触发了。

8.3.6 With Periodic Watermarks案例总结

Flink应该如何设置最大乱序时间?

- 要结合自己的业务以及数据情况进行设置。如果maxOutOfOrderness设置得太小,而自身数据发送时由于网络等原因导致乱序或者延迟太多,那么最终的结果就是会有很多单条的数据在Window中被触发,这对数据的正确性影响太大。

- 对于严重乱序的数据,需要严格统计数据最大延迟时间,这样才能保证计算的数据准确。延时设置得太小会影响数据的准确性;延时设置得太大不仅影响数据的实时性,更加会加重Flink作业的负担。对EventTime要求不是特别严格的数据,尽量不要采用EventTime方式来处理,否则有丢失数据的风险。

第 9 章
Flink 并行度详解

本章主要针对 Flink 中的并行度进行详细分析。Flink 中的并行度设置分为 4 个层面：Operator Level（算子层面）、Execution Environment Level（执行环境层面）、Client Level（客户端层面）和 System Level（系统层面）。

9.1　Flink 并行度

一个 Flink 程序由多个任务（Source、Transformation 和 Sink）组成。一个任务由多个并行实例（线程）来执行，一个任务的并行实例（线程）数目被称为该任务的并行度。

9.2　TaskManager 和 Slot

Flink 的每个 TaskManager 为集群提供 Solt。Solt 的数量通常与每个 TaskManager 节点的可用 CPU 内核数成比例，一般情况下 Slot 的数量就是每个节点的 CPU 的核数，如图 9.1 和图 9.2 所示。

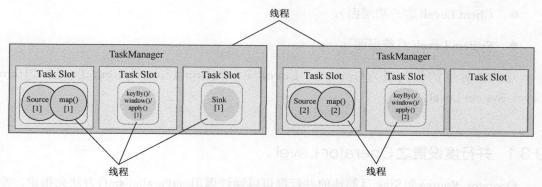

图9.1 TaskManager与Slot（1）

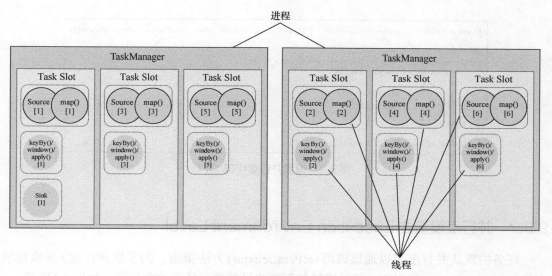

图9.2 TaskManager与Slot（2）

9.3 并行度的设置

一个任务的并行度设置可以从4个层面指定。

- Operator Level（算子层面）。
- Execution Environment Level（执行环境层面）。

- Client Level（客户端层面）。
- System Level（系统层面）。

这些并行度的优先级为Operator Level>Execution Environment Level>Client Level>System Level。

9.3.1 并行度设置之Operator Level

Operator、Source 和 Sink 目的地的并行度可以通过调用 setParallelism() 方法来指定，参考代码如图9.3所示。

```
final StreamExecutionEnvironment env = StreamExecutionEnvironment.getExecutionEnvironment();

DataStream<String> text = [...]
DataStream<Tuple2<String, Integer>> wordCounts = text
    .flatMap(new LineSplitter())
    .keyBy(0)
    .timeWindow(Time.seconds(5))
    .sum(1).setParallelism(5);

wordCounts.print();

env.execute("Word Count Example");
```

图9.3 设置并行度代码（1）

9.3.2 并行度设置之Execution Environment Level

任务的默认并行度可以通过调用 setParallelism() 方法指定。为了以并行度3来执行所有的Operator、Source 和 Sink，可以通过如下方式设置执行环境的并行度，如图9.4所示。

注意：执行环境（env）的并行度可以通过显式设置算子的并行度来重写。

```
final StreamExecutionEnvironment env = StreamExecutionEnvironment.getExecutionEnvironment();
env.setParallelism(3);

DataStream<String> text = [...]
DataStream<Tuple2<String, Integer>> wordCounts = [...]
wordCounts.print();

env.execute("Word Count Example");
```

图9.4 设置并行度代码（2）

9.3.3 并行度设置之 Client Level

并行度还可以在客户端提交 Job 到 Flink 时设定。对于 CLI 客户端，可以通过 -p 参数指定并行度。

```
./bin/flink run -p 10 WordCount-java.jar
```

这里表示把并行度设置为10。

9.3.4 并行度设置之 System Level

在系统级可以通过设置 flink-conf.yaml 文件中的 parallelism.default 属性来指定所有执行环境的默认并行度。

9.4 并行度案例分析

Flink 集群的基础环境如图 9.5 所示。Flink 集群中有 3 个 TaskManager 节点，在集群的 flink-conf.yaml 配置文件中设置 taskmanager.numberOfTaskSlots 的值为 3，这个值的大小建议和节点 CPU 的数量保持一致。

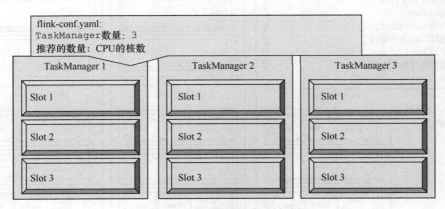

图 9.5　Flink 集群基本环境分析

此 Flink 集群中有 3 个 TaskManager 节点，每个节点有 3 个 Slot。

1. 案例1

如图9.6所示，默认情况下，WordCount任务的并行度为1，它从flink-conf.yaml文件中读取parallelism.default参数的默认值，此时只占用一个Slot。

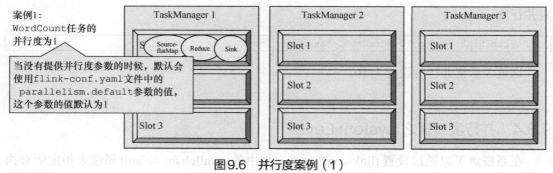

图9.6 并行度案例（1）

2. 案例2

如图9.7所示，WordCount任务的并行度为2，可以通过以下几种方式来实现，此时占用2个Slot。

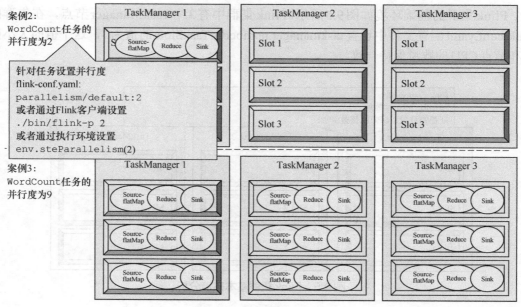

图9.7 并行度案例（2）

（1）修改 flink-conf.yaml 文件中的 parallelism.default 的值为 2。

（2）在提交任务的时候通过 bin/flink -p 2 来指定。

（3）在代码中通过 env.setParallelism(2) 来指定。

3. 案例 3

WordCount 任务的并行度为 9，可以通过以下几种方式来实现，此时占用 9 个 Slot。

（1）修改 flink-conf.yaml 文件中的 parallelism.default 的值为 9。

（2）在提交任务的时候通过 bin/flink -p 9 来指定。

（3）在代码中通过 env.setParallelism(9) 来指定。

4. 案例 4

如图 9.8 所示，WordCount 任务的并行度为 9，但是 Sink 组件的并行度为 1。

此时，Sink 组件的并行度需要在代码中通过 setParallelism(1) 来设置。

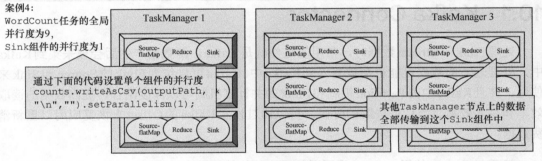

图 9.8 并行度案例（3）

第10章
Flink Kafka Connector详解

Flink提供了很多Connector组件，其中应用较广泛的就是Kafka了。本章我们主要针对Kafka Connector在Flink中的应用做详细的分析。

10.1 Kafka Connector

处理Flink的Stream数据时，常用的组件就是Kafka。源数据产生后会被采集到Kafka中，处理之后的数据可能也会被写入到Kafka中。Kafka可以作为Flink的Source和Sink来使用，并且Kafka中的Partition机制和Flink的并行度机制可以深度结合，提高数据的读取效率和写入效率。当任务失败的时候，可以通过设置Kafka的Offset来恢复应用以重新消费数据。

想要在Flink中使用Kafka，需要添加对应的依赖，因为Kafka Connector这个组件的依赖代码没有集成在Flink的核心代码中。

```
<dependency>
    <groupId>org.apache.flink</groupId>
    <artifactId>flink-connector-kafka-0.11_2.11</artifactId>
    <version>1.6.1</version>
</dependency>
```

10.2 Kafka Consumer

10.2.1 Kafka Consumer消费策略设置

Flink从Kafka中消费数据的Java代码如下。

```java
package xuwei.tech.streaming;

import org.apache.Flink.api.common.serialization.SimpleStringSchema;
import org.apache.Flink.streaming.api.DataStream.DataStreamSource;
import org.apache.Flink.streaming.api.environment.StreamExecutionEnvironment;
import org.apache.Flink.streaming.connectors.kafka.FlinkKafkaConsumer011;

import java.util.Properties;
/**
 * kafkaSource
 *
 * Created by xuwei.tech
 */
public class StreamingKafkaSource {

    public static void main(String[] args) throws Exception {
        //获取Flink的运行环境
        StreamExecutionEnvironment env = StreamExecutionEnvironment.
        getExecutionEnvironment();

        String topic = "t1";
        Properties prop = new Properties();
        prop.setProperty("bootstrap.servers","hadoop110:9092");
        prop.setProperty("group.id","con1");

        FlinkKafkaConsumer011<String> myConsumer = new FlinkKafkaConsumer011<>(topic,
        new SimpleStringSchema(), prop);

        myConsumer.setStartFromGroupOffsets();//默认消费策略

        DataStreamSource<String> text = env.addSource(myConsumer);
```

```java
            text.print().setParallelism(1);

            env.execute("StreamingKafkaSource");


    }
}
```

Scala代码如下。

```scala
package xuwei.tech.streaming

import java.util.Properties

import org.apache.Flink.api.common.serialization.SimpleStringSchema
import org.apache.Flink.contrib.streaming.state.RocksDBStateBackend
import org.apache.Flink.streaming.api.CheckpointingMode
import org.apache.Flink.streaming.api.environment.CheckpointConfig
import org.apache.Flink.streaming.api.scala.StreamExecutionEnvironment
import org.apache.Flink.streaming.connectors.kafka.FlinkKafkaConsumer011

/**
  * Created by xuwei.tech
  */
object StreamingKafkaSourceScala {

  def main(args: Array[String]): Unit = {

    val env = StreamExecutionEnvironment.getExecutionEnvironment

    //隐式转换
    import org.apache.Flink.api.scala._

    val topic = "t1"
    val prop = new Properties()
    prop.setProperty("bootstrap.servers","hadoop110:9092")
    prop.setProperty("group.id","con1")



    val myConsumer = new FlinkKafkaConsumer011[String](topic,new SimpleStringSchema(),prop)

    val text = env.addSource(myConsumer)
```

```
    text.print()
    env.execute("StreamingKafkaSourceScala ")
  }
}
```

在这里分析 Kafka Consumer 消费策略。

1. setStartFromGroupOffsets()（默认消费策略）

在上面的消费代码中，我们设置了 myConsumer.setStartFromGroupOffsets()，它是默认的消费策略，会读取上次消费者保存的 Offset 信息。如果任务是第一次启动，读取不到上次的 Offset 信息，则会根据参数 auto.offset.reset 的值来消费数据。

2. setStartFromEarliest()

从最初的数据开始进行消费，忽略存储的 Offset 信息。

3. setStartFromLatest()

从最新的数据进行消费，忽略存储的 Offset 信息。

4. setStartFromSpecificOffsets(Map<KafkaTopicPartition, Long>)

可以在代码中指定每个分区开始读取的 Offset 信息，参考代码如图 10.1 所示。

```
Map<KafkaTopicPartition, Long> specificStartOffsets = new HashMap<>();
specificStartOffsets.put(new KafkaTopicPartition("myTopic", 0), 23L);
specificStartOffsets.put(new KafkaTopicPartition("myTopic", 1), 31L);
specificStartOffsets.put(new KafkaTopicPartition("myTopic", 2), 43L);

myConsumer.setStartFromSpecificOffsets(specificStartOffsets);
```

图 10.1　指定 Topic 每个分区的 Offset 信息

10.2.2　Kafka Consumer 的容错

当 CheckPoint 机制开启的时候，Kafka Consumer 会定期把 Kafka 的 Offset 信息以及其他 Operator 的状态信息保存起来。当 Job 失败重启的时候，Flink 会从最近一次的 CheckPoint 中恢复数据，重新消费 Kafka 中的数据。

为了使用支持容错的 Kafka Consumer，需要开启 CheckPoint，可以通过下面的代码开启。

```
env.enableCheckpointing(5000); // 开启并且实现每5s CheckPoint一次
```

其中针对Checkpoint还支持以下配置。

```
//CheckPoint的相关配置
//每隔5000ms启动一个检查点（设置CheckPoint的周期）
env.enableCheckpointing(5000);
//设置模式为.EXACTLY_ONCE(默认值)，还可以设置为AT_LEAST_ONCE
env.getCheckpointConfig().setCheckpointingMode(CheckpointingMode.EXACTLY_ONCE);
//确保检查点之间有至少60000ms的间隔（checkpoint的最小间隔）
env.getCheckpointConfig().setMinPauseBetweenCheckpoints(500);
//检查点必须在1min内完成，或者被丢弃（checkPoint的超时时间）
env.getCheckpointConfig().setCheckpointTimeout(60000);
//同一时间只允许执行一个检查点
env.getCheckpointConfig().setMaxConcurrentCheckpoints(1);
//表示Flink处理程序被cancel后，会保留CheckPoint数据，以便根据实际需要恢复到指定的CheckPoint
env.getCheckpointConfig().enableExternalizedCheckpoints(CheckpointConfig.
ExternalizedCheckpointCleanup.RETAIN_ON_CANCELLATION);
//设置State存储的位置
env.setStateBackend(new RocksDBStateBackend("hdfs://hadoop100:9000/flink/
checkpoints",true));
```

10.2.3 动态加载Topic

使用Flink消费Kafka并指定Topic数据的时候，可以通过一个正则表达式来动态读取多个Topic中的数据，代码如图10.2所示。

```
final StreamExecutionEnvironment env = StreamExecutionEnvironment.getExecutionEnvironment();

Properties properties = new Properties();
properties.setProperty("bootstrap.servers", "localhost:9092");
properties.setProperty("group.id", "test");

FlinkKafkaConsumer011<String> myConsumer = new FlinkKafkaConsumer011<>(
    java.util.regex.Pattern.compile("test-topic-[0-9]"),
    new SimpleStringSchema(),
    properties);

DataStream<String> stream = env.addSource(myConsumer);
...
```

图10.2 动态加载Topic

由于主要通过Pattern.compile()来指定一个Topic的正则表达式，因此这样只要是满足这个规则的Topic的数据都可以被消费。

10.2.4 Kafka Consumer Offset 自动提交

Kafka Consumer Offset 自动提交的配置需要根据 Job 是否开启 CheckPoint 来区分。

1. CheckPoint 关闭时

可以通过下面两个参数配置。

```
enable.auto.commit: true
auto.commit.interval.ms: 1000
```

2. CheckPoint 开启时

当执行 CheckPoint 的时候才会保存 Offset，这样保证了 Kafka 的 Offset 和 CheckPoint 的状态偏移量保持一致。

可以通过此方法设置：setCommitOffsetsOnCheckpoints(true)。

这里的参数默认就是 true，表示在 CheckPoint 的时候提交 Offset，此时 Kafka 中的 Offset 自动提交机制就会被忽略。

10.3 Kafka Producer

10.3.1 Kafka Producer 的使用

Flink 在 Kafka 中生产数据的 Java 代码如下。

```
package xuwei.tech.streaming;

import org.apache.Flink.api.common.serialization.SimpleStringSchema;
import org.apache.Flink.streaming.api.DataStream.DataStreamSource;
import org.apache.Flink.streaming.api.environment.StreamExecutionEnvironment;
import org.apache.Flink.streaming.connectors.kafka.FlinkKafkaProducer011;
import org.apache.Flink.streaming.util.serialization.KeyedSerializationSchemaWrapper;

import java.util.Properties;

/**
```

```
 * kafkaSink
 * Created by xuwei.tech
 */
public class StreamingKafkaSink {

    public static void main(String[] args) throws Exception {
        //获取Flink的运行环境
        StreamExecutionEnvironment env = StreamExecutionEnvironment.
        getExecutionEnvironment();

        DataStreamSource<String> text = env.socketTextStream("hadoop100", 9001, "\n");

        String brokerList = "hadoop110:9092";
        String topic = "t1";

        Properties prop = new Properties();
        prop.setProperty("bootstrap.servers",brokerList);

        FlinkKafkaProducer011<String> myProducer = new FlinkKafkaProducer011<>
        (brokerList, topic, new SimpleStringSchema());

        text.addSink(myProducer);
        env.execute("StreamingKafkaSink");
    }
}
```

Scala代码如下。

```
package xuwei.tech.streaming

import java.util.Properties

import org.apache.Flink.api.common.serialization.SimpleStringSchema
import org.apache.Flink.streaming.api.CheckpointingMode
import org.apache.Flink.streaming.api.environment.CheckpointConfig
import org.apache.Flink.streaming.api.scala.StreamExecutionEnvironment
import org.apache.Flink.streaming.connectors.kafka.{FlinkKafkaConsumer011, FlinkKafkaProducer011}
import org.apache.Flink.streaming.util.serialization.KeyedSerializationSchemaWrapper

/**
```

```
 * Created by xuwei.tech
 */
object StreamingKafkaSinkScala {

  def main(args: Array[String]): Unit = {

    val env = StreamExecutionEnvironment.getExecutionEnvironment

    //隐式转换
    import org.apache.Flink.api.scala._

    val text = env.socketTextStream("hadoop100",9001,'\n')


    val topic = "t1"
    val prop = new Properties()
    prop.setProperty("bootstrap.servers","hadoop110:9092")

     FlinkKafkaProducer011<String> myProducer = new FlinkKafkaProducer011<>
(brokerList, topic, new SimpleStringSchema());
    env.execute("StreamingKafkaSinkScala ")
  }

}
```

10.3.2 Kafka Producer 的容错

10.3.1 节中的代码可以实现把数据写到 Kafka 中，但是不支持 Exactly-once 语义，因此不能保证数据的绝对安全性。在分析 Kafka Producer 的容错时，需要根据 Kafka 的不同版本分别进行。

1. Kafka 0.9 和 Kafka 0.10

如果 Flink 开启了 CheckPoint，则要想针对 FlinkKafkaProducer09 和 FlinkKafkaProducer010 可以提供 At-least-once 的语义，还需要配置下面两个参数。

- setLogFailuresOnly(false)。

- setFlushOnCheckpoint(true)。

注意：建议修改 Kafka 生产者的重试次数，retries 这个参数的默认值是 0，可以将其改为 3。

2. Kafka 0.11

如果Flink开启了CheckPoint,则针对FlinkKafkaProducer011就可以提供Exactly-once的语义了。

需要在使用的时候选择具体的语义,支持以下3个选项。

- Semantic.NONE。

- Semantic.AT_LEAST_ONCE(默认)。

- Semantic.EXACTLY_ONCE。

在这里我使用的Kafka是基于0.11这个版本,如果是低版本的话,有一些特性是不支持的,参考代码如下。

```
package xuwei.tech.streaming;

import org.apache.Flink.api.common.serialization.SimpleStringSchema;
import org.apache.Flink.streaming.api.CheckpointingMode;
import org.apache.Flink.streaming.api.DataStream.DataStreamSource;
import org.apache.Flink.streaming.api.environment.CheckpointConfig;
import org.apache.Flink.streaming.api.environment.StreamExecutionEnvironment;
import org.apache.Flink.streaming.connectors.kafka.FlinkKafkaProducer011;
import org.apache.Flink.streaming.util.serialization.KeyedSerializationSchemaWrapper;

import java.util.Properties;

/**
 * kafkaSink
 * Created by xuwei.tech
 */
public class StreamingKafkaSink {

    public static void main(String[] args) throws Exception {
        //获取Flink的运行环境
        StreamExecutionEnvironment env = StreamExecutionEnvironment.getExecutionEnvironment();

        //CheckPoint配置
```

```
        env.enableCheckpointing(5000);
        env.getCheckpointConfig().setCheckpointingMode(CheckpointingMode.EXACTLY_ONCE);
        env.getCheckpointConfig().setMinPauseBetweenCheckpoints(500);
        env.getCheckpointConfig().setCheckpointTimeout(60000);
        env.getCheckpointConfig().setMaxConcurrentCheckpoints(1);
        env.getCheckpointConfig().enableExternalizedCheckpoints(CheckpointConfig.
ExternalizedCheckpointCleanup.RETAIN_ON_CANCELLATION);

        //设置StateBackend

        env.setStateBackend(new RocksDBStateBackend("hdfs://hadoop100:9000/flink/
checkpoints",true));


        DataStreamSource<String> text = env.socketTextStream("hadoop100", 9001, "\n");

        String brokerList = "hadoop110:9092";
        String topic = "t1";

        Properties prop = new Properties();
        prop.setProperty("bootstrap.servers",brokerList);

        //这个构造函数不支持自定义语义
        //FlinkKafkaProducer011<String> myProducer = new FlinkKafkaProducer011<>
        //(brokerList, topic, new SimpleStringSchema());
        //注意：在使用Exactly-once语义的时候执行代码会报错，提示The transaction timeout
        //is larger than the maximum value allowed by the broker，因为Kafka服务中默认事
        //务的超时时间是15min，但是FlinkKafkaProducer011中的事务超时时间默认是1h。
        //第一种解决方案，设置FlinkKafkaProducer011中的事务超时时间
        //prop.setProperty("transaction.timeout.ms",60000*15+"");

        //第二种解决方案，修改Kafka的server.properties配置文件，设置Kafka的事务超时时间，修
        //改后需要重启Kafka集群服务

        //这个构造函数支持自定义语义，使用Exactly-once语义的Kafka Producer
        FlinkKafkaProducer011<String> myProducer = new FlinkKafkaProducer011<>(topic,
new KeyedSerializationSchemaWrapper<String>(new SimpleStringSchema()), prop,
FlinkKafkaProducer011.Semantic.EXACTLY_ONCE);
        text.addSink(myProducer);
        env.execute("StreamingKafkaSink");
    }
}
```

Scala代码如下。

```scala
package xuwei.tech.streaming

import java.util.Properties

import org.apache.Flink.api.common.serialization.SimpleStringSchema
import org.apache.Flink.streaming.api.CheckpointingMode
import org.apache.Flink.streaming.api.environment.CheckpointConfig
import org.apache.Flink.streaming.api.scala.StreamExecutionEnvironment
import org.apache.Flink.streaming.connectors.kafka.{FlinkKafkaConsumer011, FlinkKafkaProducer011}
import org.apache.Flink.streaming.util.serialization.KeyedSerializationSchemaWrapper

/**
  * Created by xuwei.tech
  */
object StreamingKafkaSinkScala {

  def main(args: Array[String]): Unit = {

    val env = StreamExecutionEnvironment.getExecutionEnvironment

    //隐式转换
    import org.apache.Flink.api.scala._

    //CheckPoint配置
    env.enableCheckpointing(5000);
    env.getCheckpointConfig.setCheckpointingMode(CheckpointingMode.EXACTLY_ONCE);
    env.getCheckpointConfig.setMinPauseBetweenCheckpoints(500);
    env.getCheckpointConfig.setCheckpointTimeout(60000);
    env.getCheckpointConfig.setMaxConcurrentCheckpoints(1);
    env.getCheckpointConfig.enableExternalizedCheckpoints(CheckpointConfig.ExternalizedCheckpointCleanup.RETAIN_ON_CANCELLATION);

    //设置StateBackend
    env.setStateBackend(new RocksDBStateBackend("hdfs://hadoop100:9000/flink/checkpoints",true));

    val text = env.socketTextStream("hadoop100",9001,'\n')
    val topic = "t1"
```

```scala
    val prop = new Properties()
    prop.setProperty("bootstrap.servers","hadoop110:9092")
    //第一种解决方案,设置FlinkKafkaProducer011中的事务超时时间
    //设置事务超时时间
    //prop.setProperty("transaction.timeout.ms",60000*15+"");

    //第二种解决方案,设置Kafka的最大事务超时时间

    //使用支持Exactly-once语义的形式
    val myProducer = new FlinkKafkaProducer011[String](topic,new KeyedSerializationSchemaWrapper
    [String](new SimpleStringSchema()), prop, FlinkKafkaProducer011.Semantic.EXACTLY_ONCE)
    text.addSink(myProducer)

    env.execute("StreamingKafkaSinkScala ")
  }
}
```

第11章
Flink实战项目开发

本章主要针对Flink的一些实战应用场景进行分析，包含架构设计和代码实现。在这里主要介绍两个应用场景：一个是实时数据清洗，也称为实时ETL；另一个是实时数据报表。

11.1 实时数据清洗（实时ETL）

11.1.1 需求分析

假设目前公司中有四五百台前端业务机器，每天产生T级别的业务日志数据。由于业务原因，我们把十几种类型的日志数据都通过一个接口进行日志记录。这些日志数据中的个别字段需要进行转换[如国家（地区）和大区之前的关系，一个大区对应多个国家（地区）码，因为大区和国家（地区）的对应关系会变动，所以日志中存储的是国家（地区）码，在具体使用的时候需要转换]，并且最好根据类型分开统计这些日志数据，这样可以提高后面计算程序的效率，因此需要对源日志数据进行转换，并且根据数据类型分开存储数据。

11.1.2 项目架构设计

实时数据清洗项目架构如图11.1所示。

11.1 实时数据清洗（实时ETL）

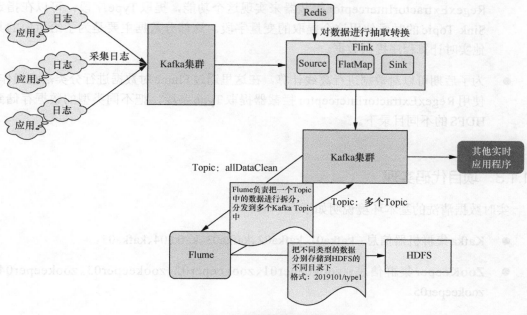

图11.1 实时数据清洗项目架构

项目架构分析如下。

- 使用Flume采集前端业务机器（应用服务器）上的日志数据，使用Exec Source监控指定文件日志数据的产生。在这里注意，需要使用tail -F，而不是tail -f，否则会导致文件重命名后无法采集新文件中的数据。

- 通过Flume把机器中的日志数据采集到Kafka中的一个Topic中，Topic的名称是allData。

- 通过Flink读取Kafka中的allData进行实时转换。首先需要对数据进行拆分，因为原始日志数据是一个嵌套JSON，我们需要把嵌套JSON进行拆分。然后再从Redis中获取最新的大区和国家（地区）码之间的映射关系，在日志中增加大区字段。在这里需要用Flink中的Connect操作把原始日志数据和Redis中的大区国家（地区）码映射关系数据关联到一起。

- 数据解析完成后，通过FlinkKafkaProducer011把数据写到Kafka中的allDataClean中。

- 这时所有类型的数据都存储到Kafka中的allDataClean了，为了减轻之后实时计算程序的压力，最好把数据拆分开，不同类型的数据存储到不同的Topic中。在日志中有一个Type字段，根据这个字段的值可以把数据分开存储。Flume使用

RegexExtractorInterceptor拦截器来实现这个功能，提取Type字段，可以在指定Sink Topic的时候使用这个提取的变量字段。这部分数据主要是为了给后面的其他实时计算程序提供数据。

- 为了后期可以对数据进行离线计算，在这里通过Flume对数据进行分类落盘操作，使用RegexExtractorInterceptor拦截器提取Type字段，把不同类型的数据存储到HDFS的不同目录下。

11.1.3 项目代码实现

实时数据清洗的基本环境说明如下。

- Kafka集群机器信息：kafka01、kafka02、kafka03、kafka04、kafka05。
- ZooKeeper集群信息：zookeeper01、zookeeper02、zookeeper03、zookeeper04、zookeeper05。
- Hadoop集群信息：hadoop01、hadoop02、hadoop03、hadoop04、hadoop05。
- Redis节点信息：redis01。

注意：针对项目中使用的相关框架的安装和部署步骤在这里不再赘述。

使用Flume将前端业务机器的日志数据采集到Kafka的allData中。

注意：这里使用的Flume版本是1.8。

Flume的配置文件file-kafka-allData.conf的内容如下。

```
#Source的名字
a1.sources = fileSource
# Channel的名字
a1.channels = memoryChannel
# Sink的名字
a1.sinks = kafkaSink

# 指定Source使用的Channel名字
a1.sources.fileSource.channels = memoryChannel
```

```
# 指定Sink需要使用的Channel的名字
a1.sinks.kafkaSink.channel = memoryChannel

# Source相关配置
a1.sources.fileSource.type = exec
a1.sources.fileSource.command = tail -F /data/log/allData.log


# Channel相关配置
a1.channels.memoryChannel.type = memory
a1.channels.memoryChannel.capacity = 1000
a1.channels.memoryChannel.transactionCapacity = 1000
a1.channels.memoryChannel.byteCapacityBufferPercentage = 20
a1.channels.memoryChannel.byteCapacity = 800000

# Sink相关配置
a1.sinks.kafkaSink.type = org.apache.flume.sink.kafka.KafkaSink
a1.sinks.kafkaSink.kafka.topic = allData
a1.sinks.kafkaSink.kafka.bootstrap.servers= kafka01:9092,kafka02:9092,kafka03:9092
a1.sinks.kafkaSink.kafka.flumeBatchSize = 20
a1.sinks.kafkaSink.kafka.producer.acks = 1
a1.sinks.kafkaSink.kafka.producer.linger.ms = 1
a1.sinks.kafkaSink.kafka.producer.compression.type = snappy
```

Flink实时解析转换程序代码的具体步骤如下。

（1）首先添加相关Maven依赖。

```xml
<dependencies>
    <dependency>
        <groupId>org.apache.flink</groupId>
        <artifactId>flink-java</artifactId>
        <version>1.6.1</version>
        <!-- provided在这里表示此依赖只在代码编译的时候使用，运行和打包的时候不使用 -->
        <!--<scope>provided</scope>-->
    </dependency>
    <dependency>
        <groupId>org.apache.flink</groupId>
        <artifactId>flink-streaming-java_2.11</artifactId>
        <version>1.6.1</version>
        <!--<scope>provided</scope>-->
    </dependency>
```

```xml
<dependency>
    <groupId>org.apache.flink</groupId>
    <artifactId>flink-scala_2.11</artifactId>
    <version>1.6.1</version>
    <!--<scope>provided</scope>-->
</dependency>
<dependency>
    <groupId>org.apache.flink</groupId>
    <artifactId>flink-streaming-scala_2.11</artifactId>
    <version>1.6.1</version>
    <!--<scope>provided</scope>-->
</dependency>

<dependency>
    <groupId>org.apache.bahir</groupId>
    <artifactId>flink-connector-redis_2.11</artifactId>
    <version>1.0</version>
</dependency>

<dependency>
    <groupId>org.apache.flink</groupId>
    <artifactId>flink-statebackend-rocksdb_2.11</artifactId>
    <version>1.6.1</version>
</dependency>

<dependency>
    <groupId>org.apache.flink</groupId>
    <artifactId>flink-connector-kafka-0.11_2.11</artifactId>
    <version>1.6.1</version>
</dependency>

<dependency>
    <groupId>org.apache.kafka</groupId>
    <artifactId>kafka-clients</artifactId>
    <version>0.11.0.3</version>
</dependency>
<!-- 日志相关依赖 -->
<dependency>
    <groupId>org.slf4j</groupId>
    <artifactId>slf4j-api</artifactId>
    <version>1.7.25</version>
```

```xml
        </dependency>

        <dependency>
            <groupId>org.slf4j</groupId>
            <artifactId>slf4j-log4j12</artifactId>
            <version>1.7.25</version>
        </dependency>
        <!-- Redis依赖 -->
        <dependency>
            <groupId>redis.clients</groupId>
            <artifactId>jedis</artifactId>
            <version>2.9.0</version>
        </dependency>
        <!-- JSON依赖 -->
        <dependency>
            <groupId>com.alibaba</groupId>
            <artifactId>fastjson</artifactId>
            <version>1.2.44</version>
        </dependency>
    </dependencies>
```

（2）因为需要从 Redis 中读取数据，所以需要自定义 RedisSource。

在自定义 RedisSource 之前，需要先在 Redis 数据库中进行数据初始化，主要是为了在 Redis 中保存国家和大区的对应关系。在后面使用的时候需要把国家和大区的对应关系组装成 Java 中的 HashMap。

在 Redis 数据库中执行以下命令。

```
hset areas AREA_US US
hset areas AREA_AR PK,KW,SA
hset areas AREA_IN IN
```

Java 代码实现如下。

```java
package xuwei.tech.source;

import org.apache.Flink.streaming.api.functions.source.SourceFunction;
import org.slf4j.Logger;
import org.slf4j.LoggerFactory;
import redis.clients.jedis.Jedis;
import redis.clients.jedis.exceptions.JedisConnectionException;
```

```java
import java.util.HashMap;
import java.util.Map;

/**
 * Created by xuwei.tech
 */
public class MyRedisSource implements SourceFunction<HashMap<String,String>> {
    private Logger logger = LoggerFactory.getLogger(MyRedisSource.class);

    private final long SLEEP_MILLION = 60000;

    private boolean isRunning = true;
    private Jedis jedis = null;

    public void run(SourceContext<HashMap<String, String>> ctx) throws Exception {
        this.jedis = new Jedis("redis01", 6379);
        //存储所有国家和大区的对应关系
        HashMap<String, String> keyValueMap = new HashMap<String, String>();
        while (isRunning){
            try{
                keyValueMap.clear();
                Map<String, String> areas = jedis.hgetAll("areas");
                for (Map.Entry<String,String> entry: areas.entrySet()) {
                    String key = entry.getKey();
                    String value = entry.getValue();
                    String[] splits = value.split(",");
                    for (String split: splits) {
                        keyValueMap.put(split,key);
                    }
                }
                if(keyValueMap.size()>0){
                    ctx.collect(keyValueMap);
                }else{
                    logger.warn("从Redis中获取的数据为空！！！");
                }
                Thread.sleep(SLEEP_MILLION);
            }catch (JedisConnectionException e){
                logger.error("Redis链接异常,重新获取链接",e.getCause());
                jedis = new Jedis("redis01", 6379);
            }catch (Exception e){
                logger.error("Source数据源异常",e.getCause());
```

```
            }
        }
    }

    public void cancel() {
        isRunning = false;
        if(jedis!=null){
            jedis.close();
        }
    }
}
```

Scala代码实现如下。

```
package xuwei.tech.source

import org.apache.Flink.streaming.api.functions.source.SourceFunction
import org.apache.Flink.streaming.api.functions.source.SourceFunction.SourceContext
import org.slf4j.LoggerFactory
import redis.clients.jedis.Jedis
import redis.clients.jedis.exceptions.JedisConnectionException

import scala.collection.mutable

/**
  * Created by xuwei.tech
  */
class MyRedisSourceScala extends SourceFunction[mutable.Map[String,String]]{

  val logger = LoggerFactory.getLogger("MyRedisSourceScala")

  val SLEEP_MILLION = 60000

  var isRunning = true
  var jedis: Jedis = _

  override def run(ctx: SourceContext[mutable.Map[String, String]]) = {
    this.jedis = new Jedis("redis01", 6379)
    //隐式转换，把Java的HashMap转为Scala的Map
    import scala.collection.JavaConversions.mapAsScalaMap
```

```scala
    //存储所有国家和大区的对应关系
    var keyValueMap = mutable.Map[String,String]()
    while (isRunning){
      try{
        keyValueMap.clear()
        keyValueMap = jedis.hgetAll("areas")

        for( key <- keyValueMap.keys.toList){
          val value = keyValueMap.get(key).get
          val splits = value.split(",")
          for(split <- splits){
            keyValueMap += (key -> split)
          }
        }

        if(keyValueMap.nonEmpty){
          ctx.collect(keyValueMap)

        }else{
          logger.warn("从Redis中获取的数据为空！！！")
        }
        Thread.sleep(SLEEP_MILLION);
      }catch {
        case e: JedisConnectionException => {
          logger.error("Redis链接异常，重新获取链接", e.getCause)
          jedis = new Jedis("redis01", 6379)
        }
        case e: Exception => {
          logger.error("Source数据源异常", e.getCause)
        }
      }
    }

  override def cancel() = {
    isRunning = false
    if(jedis!=null){
      jedis.close()
    }
  }
}
```

(3) 实现Flink数据转换任务。

注意：此处代码实现了CheckPoint和StateBackend的相关配置。

需要提前在Kafka中创建需要的Topic，创建命令如下。

```
bin/kafka-topics.sh --create --topic allData --zookeeper zookeeper01:2181--partitions 5 --replication-factor 1
bin/kafka-topics.sh --create --topic allDataClean --zookeeper zookeeper01:2181--partitions 5 --replication-factor 1
```

Java代码实现如下。

```java
package xuwei.tech;

import com.alibaba.fastjson.JSONArray;
import com.alibaba.fastjson.JSONObject;
import org.apache.Flink.api.common.serialization.SimpleStringSchema;
import org.apache.Flink.contrib.streaming.state.RocksDBStateBackend;
import org.apache.Flink.streaming.api.CheckpointingMode;
import org.apache.Flink.streaming.api.DataStream.DataStream;
import org.apache.Flink.streaming.api.DataStream.DataStreamSource;
import org.apache.Flink.streaming.api.environment.CheckpointConfig;
import org.apache.Flink.streaming.api.environment.StreamExecutionEnvironment;
import org.apache.Flink.streaming.api.functions.co.CoFlatMapFunction;
import org.apache.Flink.streaming.connectors.kafka.FlinkKafkaConsumer011;
import org.apache.Flink.streaming.connectors.kafka.FlinkKafkaProducer011;
import org.apache.Flink.streaming.util.serialization.KeyedSerializationSchemaWrapper;
import org.apache.Flink.util.Collector;
import xuwei.tech.source.MyRedisSource;

import java.util.HashMap;
import java.util.Properties;

/**
 * 数据转换清洗
 * Created by xuwei.tech
 */
public class DataClean {

    public static void main(String[] args) throws Exception{
```

```java
        StreamExecutionEnvironment env = StreamExecutionEnvironment.
getExecutionEnvironment();

        //修改并行度
        env.setParallelism(5);

        //CheckPoint配置
        env.enableCheckpointing(60000);
        env.getCheckpointConfig().setCheckpointingMode(CheckpointingMode.EXACTLY_ONCE);
        env.getCheckpointConfig().setMinPauseBetweenCheckpoints(30000);
        env.getCheckpointConfig().setCheckpointTimeout(10000);
        env.getCheckpointConfig().setMaxConcurrentCheckpoints(1);
          env.getCheckpointConfig().enableExternalizedCheckpoints(CheckpointConfig.
ExternalizedCheckpointCleanup.RETAIN_ON_CANCELLATION);

        //设置StateBackend
        //env.setStateBackend(new RocksDBStateBackend("hdfs://hadoop01:9000/flink/
checkpoints",true));

        //指定KafkaSource
        String topic = "allData";
        Properties prop = new Properties();
        prop.setProperty("bootstrap.servers","kafka01:9092,kafka02:9092");
        prop.setProperty("group.id","con1");
        FlinkKafkaConsumer011<String> myConsumer = new FlinkKafkaConsumer011<String>
(topic, new SimpleStringSchema(), prop);

        //获取Kafka中的数据
        //{"dt":"2019-01-01 11:11:11","countryCode":"US","data":[{"type":"s1","score":
0.3,"level":"A"},{"type":"s2","score":0.1,"level":"B"}]}
        DataStreamSource<String> data = env.addSource(myConsumer);

        //最新的国家码和大区的映射关系
        DataStream<HashMap<String, String>> mapData = env.addSource(new MyRedisSource()).
broadcast();//可以把数据发送到后的算子的所有并行实例中

        DataStream<String> resData = data.connect(mapData).flatMap(new CoFlatMapFunct
ion<String, HashMap<String, String>, String>() {
            //存储国家和大区的映射关系
            private HashMap<String, String> allMap = new HashMap<String, String>();
```

```java
            //flatMap1处理的是Kafka中的数据
            public void flatMap1(String value, Collector<String> out) throws Exception {
                JSONObject jsonObject = JSONObject.parseObject(value);
                String dt = jsonObject.getString("dt");
                String countryCode = jsonObject.getString("countryCode");
                //获取大区
                String area = allMap.get(countryCode);

                JSONArray jsonArray = jsonObject.getJSONArray("data");
                for (int i = 0; i < jsonArray.size(); i++) {
                    JSONObject jsonObject1 = jsonArray.getJSONObject(i);
                    jsonObject1.put("area", area);
                    jsonObject1.put("dt", dt);
                    out.collect(jsonObject1.toJSONString());
                }
            }

            //flatMap2处理的是Redis返回的Map类型的数据
            public void flatMap2(HashMap<String, String> value, Collector<String> out) throws Exception {
                this.allMap = value;
            }
        });

        String outTopic = "allDataClean";
        Properties outprop = new Properties();
        outprop.setProperty("bootstrap.servers","kafka01:9092,kafka02:9092");
        //第一种解决方案,设置FlinkKafkaProducer011中的事务超时时间
        //prop.setProperty("transaction.timeout.ms",60000*15+"");

        //第二种解决方案,设置Kafka的最大事务超时时间

        FlinkKafkaProducer011<String> myProducer = new FlinkKafkaProducer011<String>(outTopic, new KeyedSerializationSchemaWrapper<String>(new SimpleStringSchema()), outprop, FlinkKafkaProducer011.Semantic.EXACTLY_ONCE);
        resData.addSink(myProducer);

        env.execute("DataClean");
    }
}
```

Scala代码实现如下。

```scala
package xuwei.tech

import java.util.Properties

import com.alibaba.fastjson.{JSON, JSONObject}
import org.apache.Flink.api.common.serialization.SimpleStringSchema
import org.apache.Flink.contrib.streaming.state.RocksDBStateBackend
import org.apache.Flink.streaming.api.CheckpointingMode
import org.apache.Flink.streaming.api.environment.CheckpointConfig
import org.apache.Flink.streaming.api.functions.co.CoFlatMapFunction
import org.apache.Flink.streaming.api.scala.StreamExecutionEnvironment
import org.apache.Flink.streaming.connectors.kafka.{FlinkKafkaConsumer011, FlinkKafkaProducer011}
import org.apache.Flink.streaming.util.serialization.KeyedSerializationSchemaWrapper
import org.apache.Flink.util.Collector
import xuwei.tech.source.MyRedisSourceScala

import scala.collection.mutable

/**
  * Created by xuwei.tech
  */
object DataCleanScala {

  def main(args: Array[String]): Unit = {
    val env = StreamExecutionEnvironment.getExecutionEnvironment

    //修改并行度
    env.setParallelism(5)

    //checkPoint配置
    env.enableCheckpointing(60000)
    env.getCheckpointConfig.setCheckpointingMode(CheckpointingMode.EXACTLY_ONCE)
    env.getCheckpointConfig.setMinPauseBetweenCheckpoints(30000)
    env.getCheckpointConfig.setCheckpointTimeout(10000)
    env.getCheckpointConfig.setMaxConcurrentCheckpoints(1)
    env.getCheckpointConfig.enableExternalizedCheckpoints(CheckpointConfig.ExternalizedCheckpointCleanup.RETAIN_ON_CANCELLATION)

    //设置StateBackend


    //env.setStateBackend(new RocksDBStateBackend("hdfs://hadoop01:9000/flink/checkpoints",true))
```

11.1 实时数据清洗（实时ETL）

```scala
//隐式转换
import org.apache.Flink.api.scala._
val topic = "allData"
val prop = new Properties()
prop.setProperty("bootstrap.servers","kafka01:9092,kafka02:9092")
prop.setProperty("group.id","con2")

val myConsumer = new FlinkKafkaConsumer011[String]("hello",new SimpleStringSchema(),prop)
//获取Kafka中的数据
val data = env.addSource(myConsumer)

//最新的国家码和大区的映射关系
val mapData = env.addSource(new MyRedisSourceScala).broadcast //可以把数据发送到后面算子的所有并行实例中

val resData = data.connect(mapData).flatMap(new CoFlatMapFunction[String, mutable.Map[String, String], String] {

    //存储国家和大区的映射关系
    var allMap = mutable.Map[String,String]()

    override def flatMap1(value: String, out: Collector[String]): Unit = {

        val jsonObject = JSON.parseObject(value)
        val dt = jsonObject.getString("dt")
        val countryCode = jsonObject.getString("countryCode")
        //获取大区
        val area = allMap.get(countryCode)

        val jsonArray = jsonObject.getJSONArray("data")
        for (i <- 0 to jsonArray.size()-1) {
          val jsonObject1 = jsonArray.getJSONObject(i)
          jsonObject1.put("area", area)
          jsonObject1.put("dt", dt)
          out.collect(jsonObject1.toString)
        }
    }

    override def flatMap2(value: mutable.Map[String, String], out: Collector[String]): Unit = {
      this.allMap = value
```

```
            }
        })


        val outTopic = "allDataClean"
        val outprop = new Properties()
        outprop.setProperty("bootstrap.servers","kafka01:9092,kafka02:9092")
        //第一种解决方案,设置FlinkKafkaProducer011中的事务超时时间
        //prop.setProperty("transaction.timeout.ms",60000*15+"")

        //第二种解决方案,设置Kafka的最大事务超时时间

        val myProducer = new FlinkKafkaProducer011[String](outTopic, new KeyedSerializa
tionSchemaWrapper[String](new SimpleStringSchema), outprop, FlinkKafkaProducer011.
Semantic.EXACTLY_ONCE)
        resData.addSink(myProducer)


        env.execute("DataCleanScala")
    }
}
```

(4) 模拟产生测试数据的代码。

需求:模拟产生测试日志数据,直接把数据输出到Kafka的allData中。

代码实现如下。

```
package xuwei.tech.utils;

import org.apache.kafka.clients.producer.KafkaProducer;
import org.apache.kafka.clients.producer.ProducerRecord;
import org.apache.kafka.common.serialization.StringSerializer;

import java.text.SimpleDateFormat;
import java.util.Date;
import java.util.Properties;
import java.util.Random;

/**
 * Created by xuwei.tech
 */
public class kafkaProducer {
```

11.1 实时数据清洗（实时ETL）

```java
public static void main(String[] args) throws Exception{
    Properties prop = new Properties();
    //指定kafka Broker地址
    prop.put("bootstrap.servers", "kafka01:9092,kafka02:9092");
    //指定Key Value的序列化方式
    prop.put("key.serializer", StringSerializer.class.getName());
    prop.put("value.serializer", StringSerializer.class.getName());
    //指定Topic名称
    String topic = "allData";

    //创建Producer连接
    KafkaProducer<String, String> producer = new KafkaProducer<String,String>(prop);

    //{"dt":"2019-01-01 10:11:11","countryCode":"US","data":[{"type":"s1","score":0.3,"level":"A"},{"type":"s2","score":0.2,"level":"B"}]}

    //生产消息
    while(true){
        String message = "{\"dt\":\""+getCurrentTime()+"\",\"countryCode\":\""+getCountryCode()+"\",\"data\":[{\"type\":\""+getRandomType()+"\",\"score\":"+getRandomScore()+",\"level\":\""+getRandomLevel()+"\"},{\"type\":\""+getRandomType()+"\",\"score\":"+getRandomScore()+",\"level\":\""+getRandomLevel()+"\"}]}";
        System.out.println(message);
        //往Kafka的指定Topic中生产数据
        producer.send(new ProducerRecord<String, String>(topic,message));
        Thread.sleep(2000);
    }
    //关闭链接
    //producer.close();
}

public static String getCurrentTime(){
    SimpleDateFormat sdf = new SimpleDateFormat("YYYY-MM-dd HH:mm:ss");
    return sdf.format(new Date());
}

public static String getCountryCode(){
    String[] types = {"US","PK","KW","SA","IN"};
    Random random = new Random();
    int i = random.nextInt(types.length);
    return types[i];
}
```

```java
        public static String getRandomType(){
            String[] types = {"s1","s2","s3","s4","s5"};
            Random random = new Random();
            int i = random.nextInt(types.length);
            return types[i];
        }
        public static double getRandomScore(){
            double[] types = {0.3,0.2,0.1,0.5,0.8};
            Random random = new Random();
            int i = random.nextInt(types.length);
            return types[i];
        }
        public static String getRandomLevel(){
            String[] types = {"A","A+","B","C","D"};
            Random random = new Random();
            int i = random.nextInt(types.length);
            return types[i];
        }
    }
```

（5）代码运行步骤。

首先执行测试程序，持续往 Kafka 的 allData 中产生数据。然后确认 Redis 中的大区和国家码映射关系是否已经初始化。最后运行 Flink 的转换清洗程序，启动一个 Console 消费者来验证 allDataClean 中的数据。如果能正常看到大区和日期字段，就说明实时解析这部分的代码没有问题了。

（6）打 JAR 包的配置如下。

```xml
<build>
    <plugins>
        <!-- 编译插件 -->
        <plugin>
            <groupId>org.apache.maven.plugins</groupId>
            <artifactId>maven-compiler-plugin</artifactId>
            <version>3.6.0</version>
            <configuration>
                <source>1.8</source>
```

```xml
            <target>1.8</target>
            <encoding>UTF-8</encoding>
        </configuration>
    </plugin>
    <!-- Scala编译插件 -->
    <plugin>
        <groupId>net.alchim31.maven</groupId>
        <artifactId>scala-maven-plugin</artifactId>
        <version>3.1.6</version>
        <configuration>
            <scalaCompatVersion>2.11</scalaCompatVersion>
            <scalaVersion>2.11.12</scalaVersion>
            <encoding>UTF-8</encoding>
        </configuration>
        <executions>
            <execution>
                <id>compile-scala</id>
                <phase>compile</phase>
                <goals>
                    <goal>add-source</goal>
                    <goal>compile</goal>
                </goals>
            </execution>
            <execution>
                <id>test-compile-scala</id>
                <phase>test-compile</phase>
                <goals>
                    <goal>add-source</goal>
                    <goal>testCompile</goal>
                </goals>
            </execution>
        </executions>
    </plugin>
    <!-- 打JAR包插件(包含所有依赖) -->
    <plugin>
        <groupId>org.apache.maven.plugins</groupId>
        <artifactId>maven-assembly-plugin</artifactId>
        <version>2.6</version>
        <configuration>
            <descriptorRefs>
                <descriptorRef>jar-with-dependencies</descriptorRef>
            </descriptorRefs>
```

```xml
                    <archive>
                        <manifest>
                            <!-- 可以设置JAR包的入口类（可选） -->
                            <mainClass></mainClass>
                        </manifest>
                    </archive>
                </configuration>
                <executions>
                    <execution>
                        <id>make-assembly</id>
                        <phase>package</phase>
                        <goals>
                            <goal>single</goal>
                        </goals>
                    </execution>
                </executions>
            </plugin>
        </plugins>
    </build>
```

（7）封装执行脚本。

```
#!/bin/bash
# 需要在/etc/profile中配置FLINK_HOME
flink run -m yarn-cluster \
-yqu default \
-ynm DataCleanJob \
-yn 2 \
-ys 2 \
-yjm 1024 \
-ytm 1024 \
-c xuwei.tech.DataClean \
/data/soft/jars/DataClean/DataClean-1.0-SNAPSHOT-jar-with-dependencies.jar
```

完整的代码可以查阅本书的配套资源。

通过Flume对allDataClean中的数据进行解析，拆分并输出到不同的Topic（注意，这些Topic需要提前创建，或者打开Kafka中的自动创建Topic机制）中。Flume的配置文件kafka-kafka-allDataClean.conf内容如下。

```
#Source的名字
a1.sources = kafkaSource
# Channel的名字
a1.channels = fileChannel
```

```
# Sink的名字
a1.sinks = kafkaSink

# 指定Source使用的Channel中名称
a1.sources.kafkaSource.channels = fileChannel
# 指定Sink需要使用的Channel的名称
a1.sinks.kafkaSink.channel = fileChannel


# kafkaSource的相关配置
# 定义消息源类型
a1.sources.kafkaSource.type = org.apache.flume.source.kafka.KafkaSource
a1.sources.kafkaSource.batchSize = 500
a1.sources.kafkaSource.batchDurationMillis = 500
# 定义Kafka服务地址
a1.sources.kafkaSource.kafka.bootstrap.servers= kafka01:9092,kafka02:9092
# 配置消费的Kafka Topic,可以指定多个,它们之间用逗号隔开
a1.sources.kafkaSource.kafka.topics = allDataClean
# 配置消费者组的ID
a1.sources.kafkaSource.kafka.consumer.group.id = con1


# FileChannel的相关配置
# Channel类型
a1.channels.fileChannel.type = file
a1.channels.fileChannel.checkpointDir = /data/filechannle_data/checkpoint
a1.channels.fileChannel.dataDirs = /data/filechannle_data/data


# 拦截器的相关配置
# 定义拦截器
a1.sources.kafkaSource.interceptors = i1
# 设置拦截器类型
a1.sources.kafkaSource.interceptors.i1.type = regex_extractor
# 设置正则表达式,匹配指定的数据,这样设置会在数据的header中增加topic参数
a1.sources.kafkaSource.interceptors.i1.regex = "type":"(\\w+)"
a1.sources.kafkaSource.interceptors.i1.serializers = s1
a1.sources.kafkaSource.interceptors.i1.serializers.s1.name = topic


# KafkaSink的相关配置
a1.sinks.kafkaSink.type = org.apache.flume.sink.kafka.KafkaSink
```

```
a1.sinks.kafkaSink.kafka.topic = %{topic}
a1.sinks.kafkaSink.kafka.bootstrap.servers= kafka01:9092,kafka02:9092,kafka03:9092
a1.sinks.kafkaSink.kafka.flumeBatchSize = 20
a1.sinks.kafkaSink.kafka.producer.acks = 1
a1.sinks.kafkaSink.kafka.producer.linger.ms = 1
a1.sinks.kafkaSink.kafka.producer.compression.type = snappy
```

Kafka 中的数据还需要落盘存储,便于后期进行离线数据计算,这也需要使用 Flume 进行。每天的数据存储到一个目录中,并按照数据类型分别存储到不同的目录下。使用 Flume 直接从 allDataClean 中消费数据即可。Flume 的配置文件 kafka-file-allDataClean.conf 内容如下。

```
#Source的名称
a1.sources = kafkaSource
# Channel的名称
a1.channels = fileChannel
# Sink的名称
a1.sinks = hdfsSink

# 指定Source使用的Channel的名称
a1.sources.kafkaSource.channels = fileChannel
# 指定Sink需要使用的Channel的名称,注意这里是Channel
a1.sinks.hdfsSink.channel = fileChannel


# kafkaSource的相关配置
# 定义消息源类型
a1.sources.kafkaSource.type = org.apache.flume.source.kafka.KafkaSource
a1.sources.kafkaSource.batchSize = 500
a1.sources.kafkaSource.batchDurationMillis = 500
# 定义Kafka的服务地址
a1.sources.kafkaSource.kafka.bootstrap.servers= kafka01:9092,kafka02:9092
# 配置消费的Kafka Topic,可以指定多个,它们之间用逗号隔开
a1.sources.kafkaSource.kafka.topics = allDataClean
# 配置消费者组的ID
a1.sources.kafkaSource.kafka.consumer.group.id = con2



# FileChannel
# Channel类型
a1.channels.fileChannel.type = file
a1.channels.fileChannel.checkpointDir = /data/filechannle_data_2/checkpoint
a1.channels.fileChannel.dataDirs = /data/filechannle_data_2/data
```

```
# 拦截器的相关配置
# 定义拦截器
a1.sources.kafkaSource.interceptors = i1
# 设置拦截器类型
a1.sources.kafkaSource.interceptors.i1.type = regex_extractor
# 设置正则表达式,匹配指定的数据,这样设置会在数据的header中增加log_type字段
a1.sources.kafkaSource.interceptors.i1.regex = "type":"(\\w+)"
a1.sources.kafkaSource.interceptors.i1.serializers = s1
a1.sources.kafkaSource.interceptors.i1.serializers.s1.name = log_type



# hdfsSink的相关配置
a1.sinks.hdfsSink.type = hdfs
a1.sinks.hdfsSink.hdfs.path=hdfs://hadoop01:9000/data/allDataClean/%Y%m%d/%{log_type}
a1.sinks.hdfsSink.hdfs.writeFormat = Text
a1.sinks.hdfsSink.hdfs.fileType = DataStream
a1.sinks.hdfsSink.hdfs.callTimeout = 3600000

#当文件大小为52428800字节时,将临时文件滚动成一个目标文件
a1.sinks.hdfsSink.hdfs.rollSize = 52428800
#数据达到该数量的时候,将临时文件滚动成目标文件
a1.sinks.hdfsSink.hdfs.rollCount = 0
#每隔Ns将临时文件滚动成一个目标文件
a1.sinks.hdfsSink.hdfs.rollInterval = 3600

#配置前缀和后缀
a1.sinks.hdfsSink.hdfs.filePrefix=run
a1.sinks.hdfsSink.hdfs.fileSuffix=.data
```

11.2 实时数据报表

11.2.1 需求分析

实时数据报表针对一些需要实时统计的业务指标进行统计。在这里以直播平台和短视频平台中的审核指标进行统计,针对审核结果,有过审(上架)、未过审(下架)、

拉黑、推荐、上首页等类型。我们希望能够通过图表实时展现这些审核指标。数据统计的间隔满足分钟级别即可，当然精确到秒级别也是可以的，具体需要看是否能够满足业务需求。

11.2.2 项目架构设计

实时数据报表的项目架构图如图11.2所示。

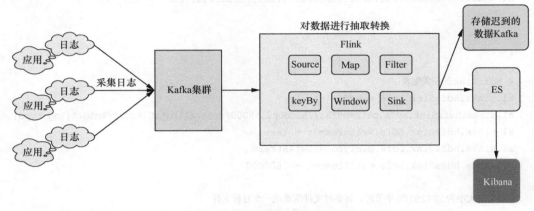

图11.2　实时数据报表项目架构

项目架构分析如下。

- 使用Flume采集前端业务机器（应用服务器）上的日志数据，使用Exec Source监控指定文件日志数据的产生。这里注意，需要使用tail -F，不能使用tail -f，否则会导致文件重命名后无法采集新文件中的数据。

- 通过Flume把机器中的日志数据采集到Kafka的一个Topic中，Topic的名称是auditLog。

- 通过Flink读取Kafka中的auditLog。首先进行过滤操作，把异常数据过滤掉。然后使用Watermark解决数据乱序的问题。因为在这里要进行分钟级别的汇总统计，所以如果想保证数据的准确性，就要处理数据乱序的问题。可以使用allowedLateness设置一个最大允许乱序时间，再使用sideOutputLateData把迟到太久的数据收集起来，方便后期排查问题。

- 再通过ElasticsearchSink把Flink最终计算的数据结果保存到ES中。

- 最后在ES后添加一个Kibana，可以很方便地查询ES中的数据并且创建一些图表。

11.2.3 项目代码实现

实时数据报表的基本环境说明如下。

- Kafka集群机器信息：kafka01、kafka02、kafka03、kafka04、kafka05。
- ZooKeeper集群信息：zookeeper01、zookeeper02、zookeeper03、zookeeper04、zookeeper05。
- Hadoop集群信息：hadoop01、hadoop02、hadoop03、hadoop04、hadoop05。
- ES集群信息：es01、es02、es03、es04、es05。

注意：针对项目中使用到的相关框架的安装和部署步骤在这里不再赘述。

使用Flume将前端业务机器的日志数据采集到Kafka的auditLog中。

Flume的配置文件file-kafka-auditLog.conf内容如下。

```
#Source的名称
a1.sources = fileSource
# Channel的名称
a1.channels = memoryChannel
# Sink的名称
a1.sinks = kafkaSink

# 指定Source使用的Channel的名称
a1.sources.fileSource.channels = memoryChannel
# 指定Sink需要使用的Channel的名称
a1.sinks.kafkaSink.channel = memoryChannel

# Source的相关配置
a1.sources.fileSource.type = exec
a1.sources.fileSource.command = tail -F /data/log/auditLog.log


# Channel的相关配置
a1.channels.memoryChannel.type = memory
a1.channels.memoryChannel.capacity = 1000
```

```
a1.channels.memoryChannel.transactionCapacity = 1000
a1.channels.memoryChannel.byteCapacityBufferPercentage = 20
a1.channels.memoryChannel.byteCapacity = 800000

# Sink相关配置
a1.sinks.kafkaSink.type = org.apache.flume.sink.kafka.KafkaSink
a1.sinks.kafkaSink.kafka.topic = auditLog
a1.sinks.kafkaSink.kafka.bootstrap.servers= kafka01:9092,kafka02:9092,kafka03:9092
a1.sinks.kafkaSink.kafka.flumeBatchSize = 20
a1.sinks.kafkaSink.kafka.producer.acks = 1
a1.sinks.kafkaSink.kafka.producer.linger.ms = 1
a1.sinks.kafkaSink.kafka.producer.compression.type = snappy
```

Flink 实时统计程序代码如下。

（1）添加相关 Maven 依赖。

```xml
<dependencies>
    <dependency>
        <groupId>org.apache.flink</groupId>
        <artifactId>flink-java</artifactId>
        <version>1.6.1</version>
        <!-- provided在这里表示此依赖只在代码编译的时候使用，运行和打包的时候不使用 -->
        <!--<scope>provided</scope>-->
    </dependency>
    <dependency>
        <groupId>org.apache.flink</groupId>
        <artifactId>flink-streaming-java_2.11</artifactId>
        <version>1.6.1</version>
        <!--<scope>provided</scope>-->
    </dependency>
    <dependency>
        <groupId>org.apache.flink</groupId>
        <artifactId>flink-scala_2.11</artifactId>
        <version>1.6.1</version>
        <!--<scope>provided</scope>-->
    </dependency>
    <dependency>
        <groupId>org.apache.flink</groupId>
        <artifactId>flink-streaming-scala_2.11</artifactId>
        <version>1.6.1</version>
        <!--<scope>provided</scope>-->
    </dependency>
```

```xml
<dependency>
    <groupId>org.apache.bahir</groupId>
    <artifactId>flink-connector-redis_2.11</artifactId>
    <version>1.0</version>
</dependency>

<dependency>
    <groupId>org.apache.flink</groupId>
    <artifactId>flink-statebackend-rocksdb_2.11</artifactId>
    <version>1.6.1</version>
</dependency>

<dependency>
    <groupId>org.apache.flink</groupId>
    <artifactId>flink-connector-kafka-0.11_2.11</artifactId>
    <version>1.6.1</version>
</dependency>

<dependency>
    <groupId>org.apache.kafka</groupId>
    <artifactId>kafka-clients</artifactId>
    <version>0.11.0.3</version>
</dependency>
<!-- 日志相关依赖 -->
<dependency>
    <groupId>org.slf4j</groupId>
    <artifactId>slf4j-api</artifactId>
    <version>1.7.25</version>
</dependency>

<dependency>
    <groupId>org.slf4j</groupId>
    <artifactId>slf4j-log4j12</artifactId>
    <version>1.7.25</version>
</dependency>
<!-- Redis依赖 -->
<dependency>
    <groupId>redis.clients</groupId>
    <artifactId>jedis</artifactId>
    <version>2.9.0</version>
</dependency>
<!-- JSON依赖 -->
```

```xml
<dependency>
    <groupId>com.alibaba</groupId>
    <artifactId>fastjson</artifactId>
    <version>1.2.44</version>
</dependency>

<!--ES依赖-->
<dependency>
    <groupId>org.apache.flink</groupId>
    <artifactId>flink-connector-elasticsearch6_2.11</artifactId>
    <version>1.6.1</version>
</dependency>
</dependencies>
```

（2）聚合统计代码实现。

首先实现自定义的聚合函数。

Java代码实现如下。

```java
package xuwei.tech.function;

import org.apache.Flink.api.java.tuple.Tuple;
import org.apache.Flink.api.java.tuple.Tuple3;
import org.apache.Flink.api.java.tuple.Tuple4;
import org.apache.Flink.streaming.api.functions.windowing.WindowFunction;
import org.apache.Flink.streaming.api.windowing.windows.TimeWindow;
import org.apache.Flink.util.Collector;

import java.text.SimpleDateFormat;
import java.util.ArrayList;
import java.util.Collections;
import java.util.Date;
import java.util.Iterator;

/**
 * 自定义聚合函数
 * Created by xuwei.tech
 */
public class MyAggFunction implements WindowFunction<Tuple3<Long, String, String>, Tuple4<String, String, String, Long>, Tuple, TimeWindow>{
    @Override
```

```java
        public void apply(Tuple tuple, TimeWindow window, Iterable<Tuple3<Long, String,
String>> input, Collector<Tuple4<String, String, String, Long>> out) throws Exception {
            //获取分组字段信息
            String type = tuple.getField(0).toString();
            String area = tuple.getField(1).toString();

            Iterator<Tuple3<Long, String, String>> it = input.iterator();

            //存储时间,这是为了获取最后一条数据的时间
            ArrayList<Long> arrayList = new ArrayList<>();

            long count = 0;
            while (it.hasNext()) {
                Tuple3<Long, String, String> next = it.next();
                arrayList.add(next.f0);
                count++;
            }

            System.err.println(Thread.currentThread().getId()+",Window触发了,数据条数:
"+count);

            //排序
            Collections.sort(arrayList);

            SimpleDateFormat sdf = new SimpleDateFormat("yyyy-MM-dd HH:mm:ss");

            String time = sdf.format(new Date(arrayList.get(arrayList.size() - 1)));

            //组装结果
            Tuple4<String, String, String, Long> res = new Tuple4<>(time, type, area, count);

            out.collect(res);
        }
    }
```

在Kafka中创建需要的Topic,命令如下。

```
bin/kafka-topics.sh --create --topic lateLog --zookeeper localhost:2181 --partitions 5 --replication-factor 1
```

下面开始实现具体聚合的代码。

```java
package xuwei.tech;

import com.alibaba.fastjson.JSON;
import com.alibaba.fastjson.JSONObject;
import org.apache.Flink.api.common.functions.FilterFunction;
import org.apache.Flink.api.common.functions.MapFunction;
import org.apache.Flink.api.common.functions.RuntimeContext;
import org.apache.Flink.api.common.serialization.SimpleStringSchema;
import org.apache.Flink.api.java.tuple.Tuple3;
import org.apache.Flink.api.java.tuple.Tuple4;
import org.apache.Flink.streaming.api.CheckpointingMode;
import org.apache.Flink.streaming.api.TimeCharacteristic;
import org.apache.Flink.streaming.api.DataStream.DataStream;
import org.apache.Flink.streaming.api.DataStream.DataStreamSource;
import org.apache.Flink.streaming.api.DataStream.SingleOutputStreamOperator;
import org.apache.Flink.streaming.api.environment.CheckpointConfig;
import org.apache.Flink.streaming.api.environment.StreamExecutionEnvironment;
import org.apache.Flink.streaming.api.windowing.assigners.TumblingEventTimeWindows;
import org.apache.Flink.streaming.api.windowing.time.Time;
import org.apache.Flink.streaming.connectors.elasticsearch.ElasticsearchSinkFunction;
import org.apache.Flink.streaming.connectors.elasticsearch.RequestIndexer;
import org.apache.Flink.streaming.connectors.elasticsearch6.ElasticsearchSink;
import org.apache.Flink.streaming.connectors.kafka.FlinkKafkaConsumer011;
import org.apache.Flink.streaming.connectors.kafka.FlinkKafkaProducer011;
import org.apache.Flink.streaming.util.serialization.KeyedSerializationSchemaWrapper;
import org.apache.Flink.util.OutputTag;
import org.apache.http.HttpHost;
import org.elasticsearch.action.index.IndexRequest;
import org.elasticsearch.client.Requests;
import org.slf4j.Logger;
import org.slf4j.LoggerFactory;
import xuwei.tech.function.MyAggFunction;
import xuwei.tech.watermark.MyWatermark;

import java.text.ParseException;
import java.text.SimpleDateFormat;
import java.util.*;

/**
 *
 * Created by xuwei.tech
 */
```

```java
public class DataReport {
    private static Logger logger = LoggerFactory.getLogger(DataReport.class);

    public static void main(String[] args) throws Exception{

        StreamExecutionEnvironment env = StreamExecutionEnvironment.getExecutionEnvironment();

        //设置并行度
        env.setParallelism(5);

        //设置使用EventTime
        env.setStreamTimeCharacteristic(TimeCharacteristic.EventTime);

        //CheckPoint配置
        env.enableCheckpointing(60000);
        env.getCheckpointConfig().setCheckpointingMode(CheckpointingMode.EXACTLY_ONCE);
        env.getCheckpointConfig().setMinPauseBetweenCheckpoints(30000);
        env.getCheckpointConfig().setCheckpointTimeout(10000);
        env.getCheckpointConfig().setMaxConcurrentCheckpoints(1);
        env.getCheckpointConfig().enableExternalizedCheckpoints(CheckpointConfig.ExternalizedCheckpointCleanup.RETAIN_ON_CANCELLATION);
        //设置StateBackend

        //env.setStateBackend(new RocksDBStateBackend("hdfs://hadoop01:9000/flink/checkpoints",true));

        /**
         * 配置kafkaSource
         */
        String topic = "auditLog";
        Properties prop = new Properties();
        prop.setProperty("bootstrap.servers","kafka01:9092,kafka02:9092");
        prop.setProperty("group.id","con1");

        FlinkKafkaConsumer011<String> myConsumer = new FlinkKafkaConsumer011<>(topic, new SimpleStringSchema(), prop);

        /**
         * 获取Kafka中的数据
```

```
         *
         * 审核数据格式
         * // {"dt":"审核时间[年月日 时分秒]","type":"审核类型","username":"审核人员姓名","area":"大区"}
         *
         */
        DataStreamSource<String> data = env.addSource(myConsumer);


        /**
         * 对数据进行清洗
         */
        DataStream<Tuple3<Long, String, String>> mapData = data.map(new MapFunction<String, Tuple3<Long, String, String>>() {
            @Override
            public Tuple3<Long, String, String> map(String line) throws Exception {
                JSONObject jsonObject = JSON.parseObject(line);

                String dt = jsonObject.getString("dt");
                long time = 0;
                try {
                    SimpleDateFormat sdf = new SimpleDateFormat("yyyy-MM-dd HH:mm:ss");
                    Date parse = sdf.parse(dt);
                    time = parse.getTime();
                } catch (ParseException e) {
                    //也可以把这个日志存储到其他介质中
                    logger.error("时间解析异常,dt:" + dt, e.getCause());
                }

                String type = jsonObject.getString("type");
                String area = jsonObject.getString("area");

                return new Tuple3<>(time, type, area);
            }
        });


        /**
         * 过滤掉异常数据
         */
        DataStream<Tuple3<Long, String, String>> filterData = mapData.filter(new FilterFunction<Tuple3<Long, String, String>>() {
            @Override
```

```java
            public boolean filter(Tuple3<Long, String, String> value) throws Exception {
                boolean flag = true;
                if (value.f0 == 0) {
                    flag = false;
                }
                return flag;
            }
        });

        //保存延迟太久的数据
        OutputTag<Tuple3<Long, String, String>> outputTag = new OutputTag<Tuple3<Long, String, String>>("late-data"){};

        /**
         * 窗口统计操作
         */
        SingleOutputStreamOperator<Tuple4<String, String, String, Long>> resultData =
filterData.assignTimestampsAndWatermarks(new MyWatermark())
                .keyBy(1, 2)
                .window(TumblingEventTimeWindows.of(Time.seconds(60)))
                .allowedLateness(Time.seconds(30))//允许迟到30s
                .sideOutputLateData(outputTag)//记录延迟太久的数据
                .apply(new MyAggFunction());


        //获取延迟太久的数据
        DataStream<Tuple3<Long, String, String>> sideOutput = resultData.getSideOutput(outputTag);

        //把延迟的数据存储到Kafka中
        String outTopic = "lateLog";
        Properties outprop = new Properties();
        outprop.setProperty("bootstrap.servers","kafka01:9092,kafka02:9092");
        outprop.setProperty("transaction.timeout.ms",60000*15+"");
        FlinkKafkaProducer011<String> myProducer = new FlinkKafkaProducer011<String>
(outTopic, new KeyedSerializationSchemaWrapper<String>(new SimpleStringSchema()),
outprop, FlinkKafkaProducer011.Semantic.EXACTLY_ONCE);

        sideOutput.map(new MapFunction<Tuple3<Long,String,String>, String>() {
            @Override
            public String map(Tuple3<Long, String, String> value) throws Exception {
```

```java
                    return value.f0+"\t"+value.f1+"\t"+value.f2;
                }
        }).addSink(myProducer);


        /**
         * 把计算的结果存储到ES中
         */
        List<HttpHost> httpHosts = new ArrayList<>();
        httpHosts.add(new HttpHost("es01", 9200, "http"));

        ElasticsearchSink.Builder<Tuple4<String, String, String, Long>> esSinkBuilder =
                new ElasticsearchSink.Builder<Tuple4<String, String, String, Long>>(
                        httpHosts,
                        new ElasticsearchSinkFunction<Tuple4<String, String, String, Long>>() {
                            public IndexRequest createIndexRequest(Tuple4<String, String, String, Long> element) {
                                Map<String, Object> json = new HashMap<>();
                                json.put("time",element.f0);
                                json.put("type",element.f1);
                                json.put("area",element.f2);
                                json.put("count",element.f3);

                                //使用time+type+area保证ID唯一
                                String id = element.f0.replace(" ","_")+"-"+element.f1+"-"+element.f2;

                                return Requests.indexRequest()
                                        .index("auditindex")
                                        .type("audittype")
                                        .id(id)
                                        .source(json);
                            }

                            @Override
                            public void process(Tuple4<String, String, String, Long> element, RuntimeContext ctx, RequestIndexer indexer) {
                                indexer.add(createIndexRequest(element));
                            }
                        }
                );
        //设置批量写数据的缓冲区大小,实际工作中这个值需要调大一些
```

```
            esSinkBuilder.setBulkFlushMaxActions(1);
            resultData.addSink(esSinkBuilder.build());


            env.execute("DataReport");

        }
    }
```

Scala代码实现如下。

```
package xuwei.tech

import java.text.{ParseException, SimpleDateFormat}
import java.util.{Date, Properties}

import com.alibaba.fastjson.JSON
import org.apache.Flink.api.common.functions.RuntimeContext
import org.apache.Flink.api.common.serialization.SimpleStringSchema
import org.apache.Flink.api.java.tuple.{Tuple, Tuple4}
import org.apache.Flink.streaming.api.environment.CheckpointConfig
import org.apache.Flink.streaming.api.functions.AssignerWithPeriodicWatermarks
import org.apache.Flink.streaming.api.scala.{OutputTag, StreamExecutionEnvironment}
import org.apache.Flink.streaming.api.scala.function.WindowFunction
import org.apache.Flink.streaming.api.watermark.Watermark
import org.apache.Flink.streaming.api.windowing.assigners.TumblingEventTimeWindows
import org.apache.Flink.streaming.api.windowing.time.Time
import org.apache.Flink.streaming.api.windowing.windows.TimeWindow
import org.apache.Flink.streaming.api.{CheckpointingMode, TimeCharacteristic}
import org.apache.Flink.streaming.connectors.elasticsearch.{ElasticsearchSinkFunction, RequestIndexer}
import org.apache.Flink.streaming.connectors.kafka.{FlinkKafkaConsumer011, FlinkKafkaProducer011}
import org.apache.Flink.streaming.util.serialization.KeyedSerializationSchemaWrapper
import org.apache.Flink.util.Collector
import org.apache.http.HttpHost
import org.elasticsearch.action.index.IndexRequest
import org.elasticsearch.client.Requests
import org.slf4j.LoggerFactory

import scala.collection.mutable.ArrayBuffer
import scala.util.Sorting
```

```scala
/**
 * Created by xuwei.tech
 */
object DataReportScala {
  val logger = LoggerFactory.getLogger("DataReportScala")

  def main(args: Array[String]): Unit = {
    val env = StreamExecutionEnvironment.getExecutionEnvironment

    //修改并行度
    env.setParallelism(5)
    env.setStreamTimeCharacteristic(TimeCharacteristic.EventTime)

    //CheckPoint 配置
    env.enableCheckpointing(60000)
    env.getCheckpointConfig.setCheckpointingMode(CheckpointingMode.EXACTLY_ONCE)
    env.getCheckpointConfig.setMinPauseBetweenCheckpoints(30000)
    env.getCheckpointConfig.setCheckpointTimeout(10000)
    env.getCheckpointConfig.setMaxConcurrentCheckpoints(1)
    env.getCheckpointConfig.enableExternalizedCheckpoints(CheckpointConfig.ExternalizedCheckpointCleanup.RETAIN_ON_CANCELLATION)

    //设置 StateBackend

    //env.setStateBackend(new RocksDBStateBackend("hdfs://hadoop01:9000/flink/checkpoints",true))

    //隐式转换
    import org.apache.Flink.api.scala._
    val topic = "auditLog"
    val prop = new Properties()
    prop.setProperty("bootstrap.servers", "kafka01:9092,kafka02:9092")
    prop.setProperty("group.id", "con2")

    val myConsumer = new FlinkKafkaConsumer011[String](topic, new SimpleStringSchema(), prop)
    //获取 Kafka 中的数据
    val data = env.addSource(myConsumer)

    //对数据进行清洗
    val mapData = data.map(line => {
      val jsonObject = JSON.parseObject(line)
```

```scala
      val dt = jsonObject.getString("dt")
      var time = 0L
      try {
        val sdf = new SimpleDateFormat("yyyy-MM-dd HH:mm:ss")
        val parse = sdf.parse(dt)
        time = parse.getTime
      } catch {
        case e: ParseException => {
          logger.error("时间解析异常,dt:" + dt, e.getCause)
        }
      }

      val type1 = jsonObject.getString("type")
      val area = jsonObject.getString("area")

      (time, type1, area)
    })

    //过滤掉异常数据
    val filterData = mapData.filter(_._1 > 0)

    //保存延迟太久的数据
    // 注意:针对Java代码需要引入org.apache.Flink.util.OutputTag
    //针对Scala代码需要引入org.apache.Flink.streaming.api.scala.OutputTag
    val outputTag = new OutputTag[Tuple3[Long, String, String]]("late-data") {}

    val resultData = filterData.assignTimestampsAndWatermarks(new AssignerWithPeriodicWatermarks
[(Long, String, String)] {
      var currentMaxTimestamp = 0L
      var maxOutOfOrderness = 10000L // 允许的最大乱序时间是10s

      override def getCurrentWatermark = new Watermark(currentMaxTimestamp - maxOutOfOrderness)

      override def extractTimestamp(element: (Long, String, String), previousElement
Timestamp: Long) = {
        val timestamp = element._1
        currentMaxTimestamp = Math.max(timestamp, currentMaxTimestamp)
        timestamp
      }
    }).keyBy(1, 2)
      .window(TumblingEventTimeWindows.of(Time.seconds(60)))
      .allowedLateness(Time.seconds(30)) //允许迟到30s
```

```scala
        .sideOutputLateData(outputTag)//收集延迟太久的数据
      .apply(new WindowFunction[Tuple3[Long, String, String], Tuple4[String, String,
String, Long], Tuple, TimeWindow] {
        override def apply(key: Tuple, window: TimeWindow, input: Iterable[(Long,
String, String)], out: Collector[Tuple4[String, String, String, Long]]) = {
          //获取分组字段信息
          val type1 = key.getField(0).toString
          val area = key.getField(1).toString
          val it = input.iterator
          //存储时间,这是为了获取最后一条数据的时间
          val arrBuf = ArrayBuffer[Long]()
          var count = 0
          while (it.hasNext) {
            val next = it.next
            arrBuf.append(next._1)
            count += 1
          }
          println(Thread.currentThread.getId + ",Window触发了,数据条数:" + count)
          //排序
          val arr = arrBuf.toArray
          Sorting.quickSort(arr)

          val sdf = new SimpleDateFormat("yyyy-MM-dd HH:mm:ss")
          val time = sdf.format(new Date(arr.last))
          //组装结果
          val res = new Tuple4[String, String, String, Long]
(time, type1, area, count)
          out.collect(res)
        }
      })

    //获取延迟太久的数据
    val sideOutput = resultData.getSideOutput[Tuple3[Long, String, String]](outputTag)

    //把延迟的数据存储到Kafka中
    val outTopic = "lateLog"
    val outprop = new Properties()
    outprop.setProperty("bootstrap.servers", "kafka01:9092,kafka02:9092")
    outprop.setProperty("transaction.timeout.ms", 60000 * 15 + "")

    val myProducer = new FlinkKafkaProducer011[String](outTopic, new KeyedSerializat
ionSchemaWrapper[String](new SimpleStringSchema()), outprop, FlinkKafkaProducer011.
Semantic.EXACTLY_ONCE)
```

```scala
      sideOutput.map(tup => tup._1 + "\t" + tup._2 + "\t" + tup._3).
addSink(myProducer)


    //把计算的结果存储到ES中
    val httpHosts = new java.util.ArrayList[HttpHost]
    httpHosts.add(new HttpHost("es01", 9200, "http"))


    val esSinkBuilder = new org.apache.Flink.streaming.connectors.elasticsearch6.
ElasticsearchSink.Builder[Tuple4[String, String, String, Long]](
      httpHosts,
      new ElasticsearchSinkFunction[Tuple4[String, String, String, Long]] {
        def createIndexRequest(element: Tuple4[String, String, String, Long]): IndexRequest = {
          val json = new java.util.HashMap[String, Any]
          json.put("time", element.f0)
          json.put("type", element.f1)
          json.put("area", element.f2)
          json.put("count", element.f3)

          val id = element.f0.replace(" ", "_") + "-" + element.f1 + "-" + element.f2

          return Requests.indexRequest()
            .index("auditindex")
            .'type'("audittype")
            .id(id)
            .source(json)
        }

        override def process(element: Tuple4[String, String, String, Long], runtimeC
ontext: RuntimeContext, requestIndexer: RequestIndexer) = {
          requestIndexer.add(createIndexRequest(element))
        }
      }
    )

    esSinkBuilder.setBulkFlushMaxActions(1)

    resultData.addSink(esSinkBuilder.build())

    env.execute("DataReportScala")
  }
}
```

（3）模拟产生测试数据的代码。

```java
package xuwei.tech.utils;

import org.apache.kafka.clients.producer.KafkaProducer;
import org.apache.kafka.clients.producer.ProducerRecord;
import org.apache.kafka.common.serialization.StringSerializer;

import java.text.SimpleDateFormat;
import java.util.Date;
import java.util.Properties;
import java.util.Random;

/**
 * 模拟产生测试数据,把日志数据直接写到auditLog中
 * Created by xuwei.tech
 */
public class kafkaProducerDataReport {

    public static void main(String[] args) throws Exception{
        Properties prop = new Properties();
        //指定Kafka Broker地址
        prop.put("bootstrap.servers", "kafka01:9092,kafka02:9092");
        //指定Key Value的序列化方式
        prop.put("key.serializer", StringSerializer.class.getName());
        prop.put("value.serializer", StringSerializer.class.getName());
        //指定Topic名称
        String topic = "auditLog";

        //创建Producer链接
        KafkaProducer<String, String> producer = new KafkaProducer<String,String>(prop);

        //{"dt":"2019-01-01 10:11:22","type":"shelf","username":"shenhe1","area":"AREA_US"}

        //生产消息
        while(true){
            String message = "{\"dt\":\""+getCurrentTime()+"\",\"type\":\""+getRandomType()+"\",\"username\":\""+getRandomUsername()+"\",\"area\":\""+getRandomArea()+"\"}";
            System.out.println(message);
            producer.send(new ProducerRecord<String, String>(topic,message));
            Thread.sleep(1000);
```

```java
        }
        //关闭链接
        //producer.close();
    }

    public static String getCurrentTime(){
        SimpleDateFormat sdf = new SimpleDateFormat("YYYY-MM-dd HH:mm:ss");
        return sdf.format(new Date());
    }

    public static String getRandomArea(){
        String[] types = {"AREA_US","AREA_AR","AREA_IN","AREA_ID"};
        Random random = new Random();
        int i = random.nextInt(types.length);
        return types[i];
    }


    public static String getRandomType(){
        String[] types = {"shelf","unshelf","black","chlid_shelf","child_unshelf"};
        Random random = new Random();
        int i = random.nextInt(types.length);
        return types[i];
    }


    public static String getRandomUsername(){
        String[] types = {"shenhe1","shenhe2","shenhe3","shenhe4","shenhe5"};
        Random random = new Random();
        int i = random.nextInt(types.length);
        return types[i];
    }

}
```

（4）代码运行步骤。

首先执行测试程序，持续在 Kafka 的 allAudit 中产生数据。然后运行 Flink 的聚合统计程序，到 ES 中查看数据。如果可以正常看到数据，就说明实时聚合统计这部分的代码没有问题了。

(5) 打 JAR 包的配置如下。

```xml
<build>
    <plugins>
        <!-- 编译插件 -->
        <plugin>
            <groupId>org.apache.maven.plugins</groupId>
            <artifactId>maven-compiler-plugin</artifactId>
            <version>3.6.0</version>
            <configuration>
                <source>1.8</source>
                <target>1.8</target>
                <encoding>UTF-8</encoding>
            </configuration>
        </plugin>
        <!-- Scala编译插件 -->
        <plugin>
            <groupId>net.alchim31.maven</groupId>
            <artifactId>scala-maven-plugin</artifactId>
            <version>3.1.6</version>
            <configuration>
                <scalaCompatVersion>2.11</scalaCompatVersion>
                <scalaVersion>2.11.12</scalaVersion>
                <encoding>UTF-8</encoding>
            </configuration>
            <executions>
                <execution>
                    <id>compile-scala</id>
                    <phase>compile</phase>
                    <goals>
                        <goal>add-source</goal>
                        <goal>compile</goal>
                    </goals>
                </execution>
                <execution>
                    <id>test-compile-scala</id>
                    <phase>test-compile</phase>
                    <goals>
                        <goal>add-source</goal>
                        <goal>testCompile</goal>
                    </goals>
                </execution>
```

```xml
            </executions>
        </plugin>
        <!-- 打JAR包插件(会包含所有依赖) -->
        <plugin>
            <groupId>org.apache.maven.plugins</groupId>
            <artifactId>maven-assembly-plugin</artifactId>
            <version>2.6</version>
            <configuration>
                <descriptorRefs>
                    <descriptorRef>jar-with-dependencies</descriptorRef>
                </descriptorRefs>
                <archive>
                    <manifest>
                        <!-- 可以设置JAR包的入口类(可选) -->
                        <mainClass></mainClass>
                    </manifest>
                </archive>
            </configuration>
            <executions>
                <execution>
                    <id>make-assembly</id>
                    <phase>package</phase>
                    <goals>
                        <goal>single</goal>
                    </goals>
                </execution>
            </executions>
        </plugin>
    </plugins>
</build>
```

(6) 封装执行脚本。

```bash
#!/bin/bash
flink run -m yarn-cluster \
-yqu default \
-ynm DataReportJob \
-yn 2 \
-ys 2 \
-yjm 1024 \
-ytm 1024 \
-c xuwei.tech.DataReport \
/data/soft/jars/DataReport/DataReport-1.0-SNAPSHOT-jar-with-dependencies.jar
```

完整的代码可以在本书的配套资源中查阅。

在 Kibana 中配置 ES 中的 Index 信息，根据业务需求创建对应的图表（折线图、饼图等）。效果如图 11.3 所示。

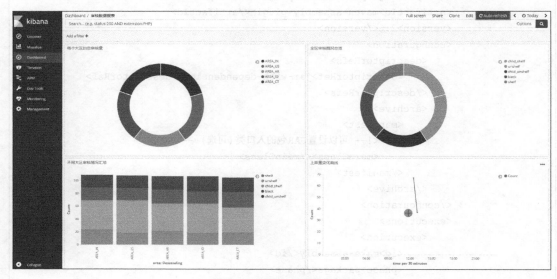

图 11.3　Kibana 展现效果